Untangling the Double Helix

DNA ENTANGLEMENT AND THE ACTION OF THE DNA TOPOISOMERASES

Related titles from Cold Spring Harbor Laboratory Press

Binding and Kinetics for Molecular Biologists
Genes & Signals
A Genetic Switch: Phage Lambda Revisited, 3rd edition
Molecular Biology of the Gene, 6th edition

Untangling the Double Helix

DNA ENTANGLEMENT AND THE ACTION OF THE DNA TOPOISOMERASES

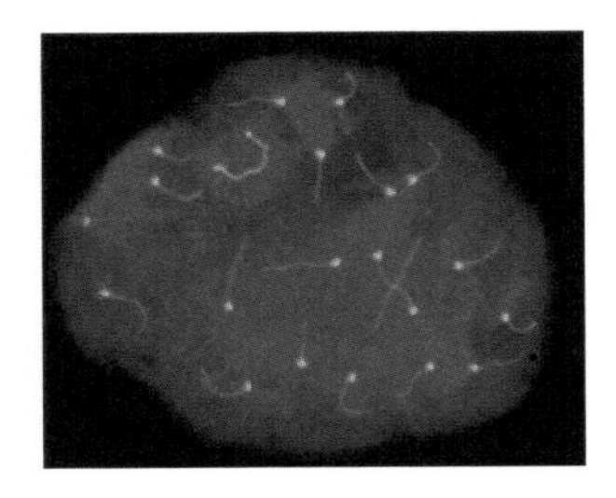

James C. Wang

Mallinckrodt Professor of Biochemistry and Molecular Biology, Emeritus, Harvard University

COLD SPRING HARBOR LABORATORY PRESS
Cold Spring Harbor, New York • www.cshlpress.com

Untangling the Double Helix
DNA Entanglement and the Action of the DNA Topoisomerases

Printed in the United States of America

Publisher	John Inglis
Acquisition Editor	Alexander Gann
Director of Development, Marketing, & Sales	Jan Argentine
Developmental Editor	Kaaren Janssen
Project Coordinator	Mary Cozza
Permissions Coordinator	Carol Brown
Production Editor	Kathleen Bubbeo
Desktop Editor	Lauren Heller
Production Manager	Denise Weiss
Book Marketing Manager	Ingrid Benirschke
Sales Account Managers	Jane Carter and Elizabeth Powers
Cover Design	Ed Atkeson

Front cover (for paperback edition only) and title page artwork: Prominent association of DNA topoisomerase IIIβ with the X-Y sex chromosome pair in a pachytene stage mouse testis cell. Spreads of chromosomes were treated with a blue fluorescent dye DAPI and different antibodies that specifically mark DNA topoisomerase IIIβ, the centromeres, and the proteinaceous axes of the chromosomes. In the fluorescence micrograph shown, the centromeres and chromosomal axes are marked in pale blue and DNA topoisomerase IIIβ is marked in pink by the use of fluorescence-labeled secondary antibodies targeting the various specific antibodies. The particular pattern of localization of DNA topoisomerase IIIβ is similar to that exhibited by several proteins known to be involved in recombination repair, including Rad50 and Rad51. See Kwan, K.Y., Moens, P.B., and Wang, J.C. 2003. *Proc. Natl. Acad. Sci.* **100:** 2526–2531 for further details. The picture shown here is a different rendition of the one reported in Fig. 3C of the reference cited.

Library of Congress Cataloging-in-Publication Data

Wang, James C.
Untangling the double helix: DNA entanglement and the action of the DNA topoisomerases / James C. Wang.
p. cm.
ISBN 978-0-87969-863-8 (cloth : alk. paper) -- ISBN 978-0-87969-879-9 (pbk. : alk. paper)
1. DNA topoisomerases--Popular works. I. Title.
QP616.D56.W36 2009
572'.786--dc22

2008054799

10 9 8 7 6 5 4 3 2 1

All Cold Spring Harbor Laboratory Press publications may be ordered directly from Cold Spring Harbor Laboratory Press, 500 Sunnyside Blvd., Woodbury, New York 11797-2924. Phone: 1-800-843-4388 in the Continental U.S. and Canada. All other locations: (516) 422-4100. FAX: (516) 422-4097. E-mail: cshpress@cshl.edu. For a complete catalog of all Cold Spring Harbor Laboratory Press publications, visit our World Wide Web site http://www.cshlpress.com/.

To Sophia, Janice, and Jessica

Contents

Preface

THIS BOOK IS ABOUT THE ENTANGLEMENT PROBLEM OF DNA inside cells and the elegant solutions Nature has come up with. The DNA entanglement problem is closely tied to the twisting of its two strands around each other and to its great length inside a cell. Initially, discussions on this problem centered on the separation of an intertwined pair of strands during replication of the double helix to endow each progeny DNA one strand of its parent, but it gradually became apparent that untwining the parental strands is but one manifestation of the entanglement problem. If a human cell nucleus containing 46 chromosomes is enlarged to the size of a basketball, the total length of DNA in the chromosomes would span 150 miles! Ingenious designs are thus needed to avoid entanglement of the 46 bundles of DNA into a hopeless mess and to remedy such a situation when the inevitable actually happens. Even in a cell not actively involved in the process of replicating its DNA, many large molecular machineries, including those busily transcribing the genetic information encoded in DNA, track along the DNA helical cables—sometimes tumbling and twirling as they follow the helical path, and other times pulling the DNA through them and forcing the double helix to twist and writhe in space. The twisting and writhing of a DNA would in turn greatly strain the threadlike molecule and perturb cellular peace and tranquility, if solutions were not found to avoid such chaos.

And fascinating solutions Nature has indeed developed. A family of enzymes, called the DNA topoisomerases, long ago evolved with DNA. These enzymes work by transiently breaking one or a pair of strands in a DNA double helix; *transiently* because the broken strand or strands are subsequently rejoined by the same enzyme that did the breakage, but in between the breakage and rejoining events another DNA strand or double helix can pass through the temporarily opened gate in a DNA strand or double helix. Thus DNA topoisomerases permit interpenetration of DNA strands and double helices but leave no chemical record of their actions on the DNA strands. The DNA strand passage events mediated by a DNA topoisomerase can be readily detected, however, by changes in the topology of DNA rings—a pair of linked DNA rings can come apart, for example, or vice versa, by passing one ring through an enzyme-mediated transient break in the other. It is the interconversion between different topological forms of DNA rings that led to the naming of these enzymes as the DNA topoisomerases, but their actions on intracellular DNA are entirely general and are by no means restricted to ring-shaped DNAs.

Throughout this book, I have tried to present the subject in a way that avoids, as much as possible, prior knowledge in chemistry, physics, or mathematics, while not overly simplifying the subject. I also tried to convey to the readers the historical development of the field—the backdrop, how some of the issues came up, how the basic concepts evolved, and how the ideas were experimentally tested, etc. The primary motivation for doing so is that, despite the gradual recognition of the significance of the DNA entanglement problem in nearly all cellular transactions of DNA, and the finding that many antimicrobial and anticancer therapeutics target the DNA topoisomerases, few articles have been written for the general readership, and many outside the field still consider the different manifestations of the DNA entanglement problem and the DNA topoisomerases exotic topics for the specialists. I hope this volume will serve as an introduction to the uninitiated and to students who are entering the field, and that a single volume covering a broad spectrum of topics may also be of some use to those who are pushing the frontiers of this field.

The book is divided into ten chapters. The first is concerned with the historic backdrop of the DNA entanglement problem during replication, and the second contains a historical account of the first encounter with an activity that alters the topology of a DNA ring. These background chapters are followed by three chapters describing the three subfamilies of the DNA topoisomerases—type IA, IB and II—each of which shares common mechanistic features yet has its own distinctive characteristics. These chapters are followed by one about the use of "single-molecule" approaches in the study of various DNA topoisomerases, in which individual molecules rather than bulk populations of molecules are examined. The next three chapters deal with the manifestations of the DNA entanglement problem and the cellular roles of the DNA topoisomerases, and the final chapter is about DNA topoisomerases as drug targets. There are also two appendices that describe some of the more chemical aspects of DNA topoisomerase-catalyzed breakage and rejoining of DNA strands.

In each chapter I have provided some of the key references, but rapid developments in the various areas have made it impossible to be inclusive or even comprehensive. I have been very fortunate in witnessing developments in this field since its very beginning four decades ago, and writing this volume brought back fond memories. I benefited greatly from discussions with many colleagues and friends during this endeavor, and several of them also kindly read various parts of the draft and provided me with invaluable suggestions. I wish to thank in particular James Berger, Carlos

Bustamante, Jim Champoux, Nancy Crisona, Patrick Forterre, Tao-Shih Hsieh, Karla Kirkegaard, Leroy Liu, Matt Melselson, Kyoshi Mizuuchi, John Nitiss, Richard Sinden, and Yuk-Ching Tse-Dinh. The excellent artwork accompanying the text was rendered by the staff of Precision Graphics. I am also grateful to Alex Gann, Kaaren Janssen, Mary Cozza, Kathleen Bubbeo, and their colleagues at the Cold Spring Harbor Laboratory Press; without their diligence, this volume would not have materialized.

James C. Wang
August 2008
Bellevue, Washington

CHAPTER 1

An Insuperable Problem

"I am willing to bet that the plectonemic coiling of the chains in your structure is radically wrong..."

Max Delbrück to James D. Watson, May 12, 1953

FEW PUBLICATIONS OF THE 20TH CENTURY can match the brevity and impact of the April 25, 1953 paper on DNA, the genetic material of all living organisms on Earth. Written by James D. Watson and Francis H.C. Crick (1916–2004), the one-page paper, published in the journal *Nature*, described a unique structure for DNA. Watson, then only 23 years old, had received his Ph.D. a little more than 2 years earlier; Crick, although almost 37, had yet to earn his doctoral degree because of his long years at the British Admiralty before embarking on graduate studies.[1] Their structure of DNA, later termed the double helix or the Watson–Crick structure, was to become one of the most celebrated images in science books and the medical literature, as well as in boardroom brochures and the news media.

Numerous pictorial renditions of this structure of DNA have since appeared, including two by the eminent Spanish surrealist painter Salvador Dali (1904–1989). However, a simple hand-drawn sketch in the original paper by Watson and Crick (Fig. 1-1a)[2] nicely captures the essential features of the double helix. In this depiction, the DNA molecule is shown to possess two separate strands, with the backbone

1. Crick thought that his late start had been a blessing in disguise. He wrote in his autobiography *What Mad Pursuit* (1988. Basic Books, Inc., New York, p. 16): "By the time most scientists have reached age thirty they are trapped by their own expertise. They have invested so much effort in one particular field that it is often extremely difficult, at that time in their careers, to make a radical change. I, on the other hand, knew nothing, except for a basic training in somewhat old-fashioned physics and mathematics and an ability to turn my hand to new things." Watson also remarked on several occasions that the mediocre work of his Ph.D. thesis had actually helped him, because it left him very little to chew on and forced him to move in a new direction.

2. Drawing done by Odile Crick, an artist and Francis Crick's wife.

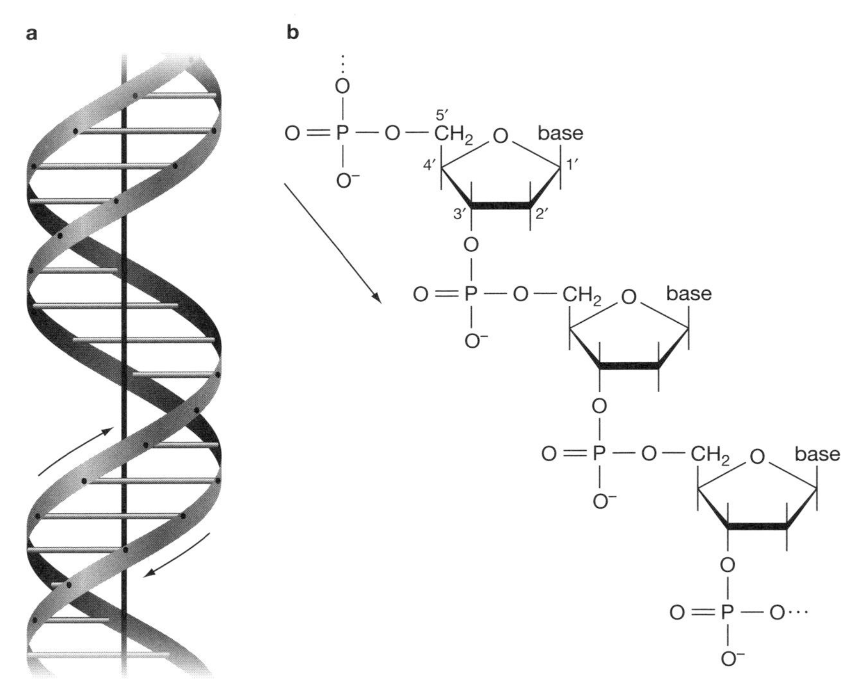

Figure 1-1. The DNA double helix. (*a*) In the double helix model of DNA proposed by James D. Watson and Francis H.C. Crick in 1953, the ribbons represent the sugar–phosphate backbone chains of the two strands of the DNA molecule, each following a right-handed helical path about a common central axis indicated by the vertical line. The chains run in opposite directions, as indicated by the two arrows. The horizontal bars represent hydrogen-bonded base pairs (bp), each of which lies in a plane perpendicular to the vertical axis. See the text for additional details. (Redrawn, by permission of Macmillan Publishers Ltd., from Watson J.D. and Crick F.H.C. 1953. *Nature* **171:** 964–967.) (*b*) The arrangement of atoms in each of the two backbone strands of DNA is shown in this three-nucleotide stretch of a strand. The numbering system for the pentose ring, the five-carbon sugar moiety, is shown for the topmost nucleotide; the five carbon atoms in the ring are numbered 1′ through 5′. The lines in the structure represent covalent bonds with the two adjoining atoms sharing one or more pairs of electrons to create the bond. The single and double lines represent, respectively, the sharing of one pair of electrons (a single bond) and the sharing of two pairs of electrons (a double bond). The oxygen (O), phosphorus (P), and carbon (C) atoms normally form two, five, and four bonds, respectively, with their neighboring atoms (counting each double bond as two bonds), whereas the hydrogen (H) atom forms only one bond with a neighboring atom. To avoid the overclustering of a diagram, the H atoms attached to a carbon atom are often omitted. For simplicity, the four carbon atoms (1′–4′) and the O atom in each sugar moiety are shown as lying in a flat plane; in its actual structure, the five-membered pentose ring is puckered, with the 2′ carbon atom of the ring projecting upward, in the same direction as the base attached to the ring. Each phosphate moiety (PO_4^-) in the backbone strand of the DNA is shown as carrying a negative charge (–), but such negative charges on the backbone chains are usually shielded by positively charged "counterions" (not shown). The arrow to the left of the sketch indicates the polarity of the backbone strand of the DNA in the 5′ to 3′ direction.

With all the accolades bestowed on DNA in the decades since Watson and Crick's 1953 paper, its humble earlier history has been all but forgotten. Thus, for example, the name guanine was derived from the word guano, in which this base was initially identified. DNA itself was first isolated in 1869, by the young Swiss physician Johann Friedrich Miescher (1844–1895), from excrements of wounds that he extracted from discarded surgical wound bandages that he obtained from a local clinic.

of each strand represented by a narrow ribbon. The two ribbons coil around a common axis resembling the side rails in a twisted ladder. Chemically, each strand of the helix is a chain of alternating sugar (deoxyribose, the "D" in DNA) and phosphate moieties. Attached to each sugar moiety is a substance known as a heterocyclic "base," which consists of atoms of carbon, nitrogen, oxygen, and hydrogen. The four different kinds of DNA bases are adenine, cytosine, guanine, and thymine, represented, respectively, by the single letters A, C, G, and T.[3]

In chemical terms, DNA is a polymer of building blocks called nucleotides, each of which consists of a deoxyribose sugar moiety, a phosphate moiety, and a base—A, C, G, or T—attached to the sugar (see Fig. 1-1b). How the sugar, phosphate, and base components are arranged in a nucleotide, and the way in which the nucleotides are linked together to form a DNA strand, had been elucidated by 1953, mainly through the efforts of the chemists Albrecht Kossel (1853–1927), Phoebus Levine (1869–1940), and Alexander Todd (1907–1997). The Watson–Crick structure specifically addressed the way in which the two strands are arranged in space.

A key feature of the DNA double helix is that its two separate strands are held together by the pairing of bases via the formation of hydrogen bonds (H-bonds). In the Watson–Crick structure, pairing of the bases follows a particular scheme: The base A specifically pairs with the base T (and vice versa) through the formation of two hydrogen bonds between them, and the base G pairs with the base C (and vice versa) through the formation of three hydrogen bonds.[4] This pairing scheme allows the different base pairs to assume a nearly uniform geometry throughout the DNA molecule (Fig. 1-2). In the illustration shown in Figure 1-1a, the hydrogen-bonded base pairs are represented as rods connecting the two backbone strands of the DNA molecule. All of the rods are of the same length because the overall shape of an A–T pair is about the same as that of a C–G pair (Fig. 1-2). Be-

3. The four bases present in DNA are sometimes modified with additional chemical groups; such modifications serve important biological functions, but these subjects are outside the scope of our discussion here.

4. In the original Watson–Crick structure of DNA, both A–T and G–C pairs were assumed to have two H-bonds per pair; it was Linus Pauling (1901–1994) who correctly noted that in a G–C pair, three H-bonds are formed between the two bases.

In the biological molecules, the formation of a hydrogen bond often involves a hydrogen (H) atom that is covalently bound either to a nitrogen (N) or an oxygen (O) atom. Atoms like N or O have a high "electronegativity," or tendency to attract and hold tightly the pair of electrons that join it to an H atom. As a consequence, an H atom that is covalently bound to an N or O atom becomes electron deficient. This deficiency in turn favors the interaction of the H atom with another, relatively electron-rich N or O atom. An H-bond is commonly represented by a dotted line, as

—N—H···N— or —O—H···O— or —O—H···N—

in which the solid lines represent the regular covalent bonds between atoms. The strength of a hydrogen bond is only a fraction of that of a typical covalent bond, such as the bond between two hydrogen atoms in a molecule of hydrogen (H—H), or that of the bonds between hydrogen and oxygen in a molecule of water (H—O—H). The presence of a large number of hydrogen bonds between two macromolecules, such as the two strands of the DNA double helix, often contributes very significantly to their association with one another.

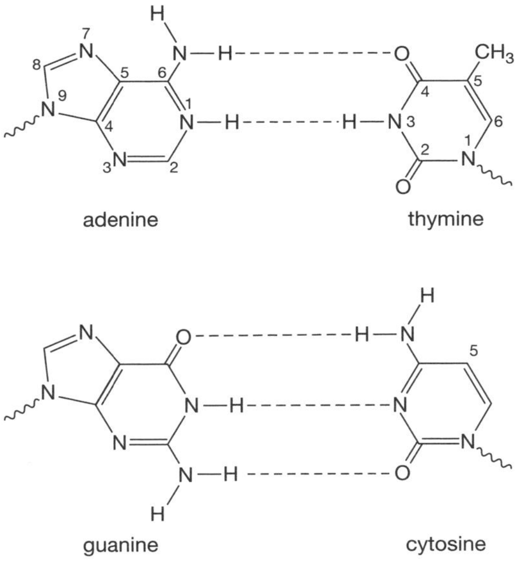

Figure 1-2. Schematic representations of an adenine–thymine (AT) base pair (*top*) and a guanine–cytosine (GC) base pair (*bottom*). Covalent bonds are represented by continuous lines and H-bonds by dotted lines; the wiggly line in each base indicates the position at which the 1′ carbon atom (C1′) of the sugar moiety is linked to the nitrogen atom of the base. Many of the H atoms are omitted in these sketches. The numbering systems for the bases A and T are shown in the *top* diagram. The atoms in G, in the *bottom* diagram, are numbered as in A in the *top* diagram (G and A belong to the class of bases known as purines), and the atoms in T (*top* diagram) are numbered as in C (*bottom* diagram) (T and C belong to the class of bases known as pyrimidines). In the *bottom* diagram, only the numbering of the C5 atom is indicated; this position is sometimes modified, for example, by methylation. The geometries of the base pairs in both the *top* and *bottom* diagrams are drawn according to the Watson–Crick model. It is significant that the overall shapes of the A–T and G–C pairs are very similar; this similarity allows the base pairs to be vertically "stacked" in an ordered way, linking the two strands of the DNA so as to form a smooth double helix. It is now known that there are significant variations in the geometry of a base pair, depending on its type and the types of its neighboring base pairs; consequently, the structure of a typical DNA molecule is not the evenly regular double helix depicted in the idealized Watson–Crick model.

cause of the asymmetric structure of a nucleotide (see Fig. 1–1b), there is a built-in polarity when the nucleotides are linked together to form one DNA strand. In the Watson–Crick double helix, the two strands run in opposite directions and are thus termed "antiparallel" as is shown in Figure 1-1a.

The discovery that base pairing holds the two strands of a DNA molecule together in the double helix structure electrified the scientific world: The Watson–Crick scheme immediately suggested a way of passing genetic information from one generation of an organism to the next. It was clear that the strands or backbones of DNA molecules, composed of an invariant and rather monotonous alteration of sugar and phosphate groups, could not possibly code for the genetic information that distinguishes one organism from another. But, different sequences of the four bases A, C, G, and T along a DNA backbone could spell out different genetic blueprints. In other words, the genetic blueprints in all forms of life could be determined by the "language" created by varying arrangements of the four letters A, G, C, and T. And, because the Watson–Crick pairing scheme specifies a one-to-one correspondence between bases on the two antiparallel strands, a sequence of 5′·····AATGCCTTA·····3′ in one strand (in which the symbols 5′ and 3′ at the ends of the sequence specify the polarity of the strand[5]) requires that the sequence of the other strand must be 3′·····TTACGGAAT·····5′. The two strands in the DNA double helix are thus not identical; rather they are "complementary" to each other.

The concept of complementarity immediately suggested that a parent DNA molecule could be duplicated by copying each strand in accordance with the same Watson–Crick scheme base-pairing: Each strand of the pair of parental strands in the duplex could thus acquire a newly synthesized complementary strand to yield two identical double helices.

Another key feature of the Watson–Crick double helix is that both strands follow a right-handed helical path around a central axis, owing to the orderly "stacking" of the planar base pairs, each of which is rotated 36° counterclockwise relative to the base pair below it. It is this very feature that makes DNA a "double helix." It is also easy to see, in the helical structure depicted in Figure 1-1a, that the two strands revolve around each other. Structures with intertwined lines are termed *plectonemic*, whereas multistranded structures in which the strands do not intertwine are called *paranemic*—terms first used in 1941 by C.L. Huskins (1897–1953) in a discussion of chromosome structures. Although many who read the 1953 Watson and Crick paper were awestruck by the principle of complementarity and its implication in passing genetic information from one generation to the next, others were skeptical about the plectonemic nature of the double-helix structure. Among the doubters was Max Delbrück (1906–1981).

5. The numbers 5′ and 3′ refer to the particular carbon atoms in the sugar moiety (see Fig. 1-1b). The notation 5′·····AATGCCTTA·····3′ specifies that the sequence of nucleotides runs in the 5′ to 3′ direction.

A physicist by training, and in his later years often referred to as the "father of molecular biology," Delbrück was in 1953 already a towering figure. The youthful Watson was in close correspondence with Delbrück both before and after the publication of the DNA double-helix paper. Delbrück was never known for mincing words. In a letter to Watson written on May 12, 1953—a month after the double-helix paper appeared—he was characteristically blunt, saying "I am willing to bet that the plectonemic coiling of the chains in your structure is radically wrong, because (1) The difficulties of untangling the chains do seem, after all, insuperable to me. (2) The X-ray data suggest only coiling but not specifically your kind of coiling."[6]

Interpretation of the DNA X-ray data, then only available for oriented DNA fibers pulled from a concentrated DNA solution, was not as straightforward as is usually perceived. As late as 1979 there were still those who argued that the famous X-ray photograph of the DNA helix could be interpreted in terms of a paranemic structure, in which two side-by-side strands were not intertwined.[7] Because the X-ray data available in 1953 could not really "prove" a particular structure for DNA, Delbrück's objection to the double-helix structure, stemming from his concern about the "untangling problem," was justified. But what exactly was this problem?

THE UNTANGLEMENT PROBLEM

Delbrück envisioned that the scheme of semiconservative replication of DNA—so designated because one-half of the progeny duplex was inherited and the other half was newly synthesized—would convert a pair of intertwined strands in the parent DNA to a pair of intertwined DNA double helices (Fig. 1-3). Unless the pair of intertwined

6. In addressing the potential problem of coiling and uncoiling of chromosomes during mitosis, Watson and Crick had written earlier: "Although it is difficult at the moment to see how these processes occur without everything getting tangled, we do not feel that this objection will be *insuperable*" (1953. *Nature* **171:** 964–967). Delbrück's choice of the adjective *"insuperable"* was therefore likely deliberate. The X-ray data to which Delbrück referred was that of Rosalind E. Franklin (1920–1958), who first obtained the X-ray diffraction patterns of ordered bundles of DNA fibers.

7. In 1976, two groups of researchers had proposed DNA structures in which two antiparallel chains of nucleotides alternate in 5-bp segments of left-handed and right-handed twists. These and additional concepts favoring a *paranemic* "side-by-side" double-stranded DNA, as well as fiery advocacy by a noted mathematician William Pohl (1938–2000), had sufficiently alarmed a no less distinguished figure than Francis Crick. Crick came to the defense of the plectonemic double-helix structure, and with colleagues published a summary of the arguments for the plectonemic structure, based on the properties of DNA rings with intact strands (Crick, F.H.C., et al. 1979. *J. Mol. Biol.* **129:** 449–461). Even after that retort, however, the champions of the side-by-side model stuck to their guns. A.G. Rodley and his collaborators in New Zealand made detailed calculations to show that their paranemic side-by-side model could fit the X-ray diffraction data for DNA fibers as well as, if not better than, could the Watson–Crick structure itself. Rodley and colleagues' results were circulated to many of those interested in DNA structure, but by then their cause was already lost. Decades after the 1979 debate, a few still remained skeptical that the double helix was the biologically important DNA structure within cells (see, e.g., the review by Delmonte, C.S. and Mann, L.R.B. 2003. *Curr. Sci.* **85:** 1564–1570).

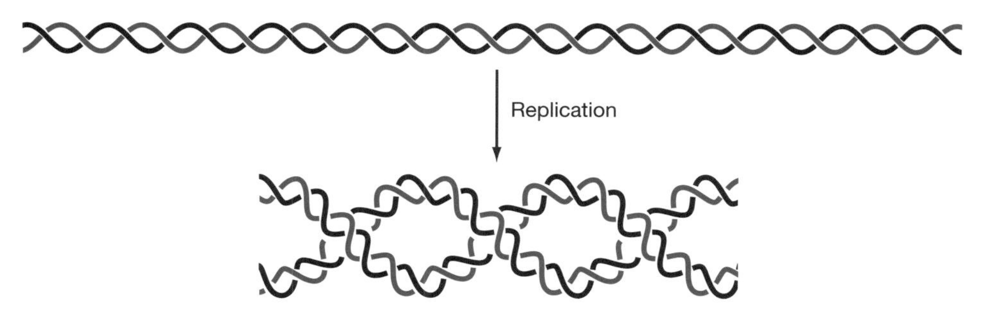

Figure 1-3. The formation of a pair of intertwined DNA double helices following the replication of a parental DNA double helix, as envisioned by Max Delbrück. Delbrück reasoned that, because of the right-handed intertwining of the two strands in a DNA molecule (*top* diagram), upon duplication of the strands, the pair of progeny DNA double helices would also be intertwined in a right-handed way (*bottom* diagram). If the two parental DNA strands did not uncoil at all as they replicated, then as the Watson–Crick double-helix model suggests, the two progeny double helices would revolve about each other approximately once for every 10 base pairs; the progeny double helices shown here are not nearly as tightly intertwined as that.

progeny helices could be completely untangled, it would be impossible for them to separate when a cell divided. Improper segregation of newly replicated DNA into two progeny cells would have disastrous consequences, because the faithful passage of genetic information from one parent cell to two progeny cells would be impaired.

In principle, the conversion of intertwined parental strands of DNA to intertwined progeny double helices could be avoided by proper rotational motions of the double helices of the parent and progeny DNA molecules around their respective axes, as illustrated in Figure 1-4.[8] In the 1950s there was little information about the sizes of DNA molecules inside a cell. Later, however, it would become clear that a DNA molecule inside a cell can be very long, and that an entire chromosome consists of a single unbroken DNA molecule that may contain huge numbers of nucleotides. The largest chromosome in a human cell, for example, contains about a quarter of a billion base pairs, and if it were extended from one end to the other, would span a length of 8.5 cm. This is an enormous length relative to the dimension of a cell nucleus, which is only about several micrometers in diameter. If a nucleus were to be enlarged to the size of a basketball, an 8.5-cm-long chromosome would become 5 kilometers or about 3 miles long! Once the length of a DNA molecule inside a cell became known,

8. Watson and Crick recognized the advantage of directly untwisting the unreplicated DNA double helix to separate it into two side-by-side strands to allow replication to occur (1953. *Cold Spring Harb. Symp. Quant. Biol.* **18:** 123–131). Others also pointed out that if a DNA replication fork is represented by a Y, its vertical stem representing the unreplicated parental portion and the two arms of the Y the growing progeny duplexes, then rotating each part of this DNA "trio" around its respective helical axis would shorten the stem and lengthen the arms, without untanglement of the arms (see, e.g., Levinthal, C. and Crane, H.R. 1956. *Proc. Natl. Acad. Sci.* **42:** 436–438).

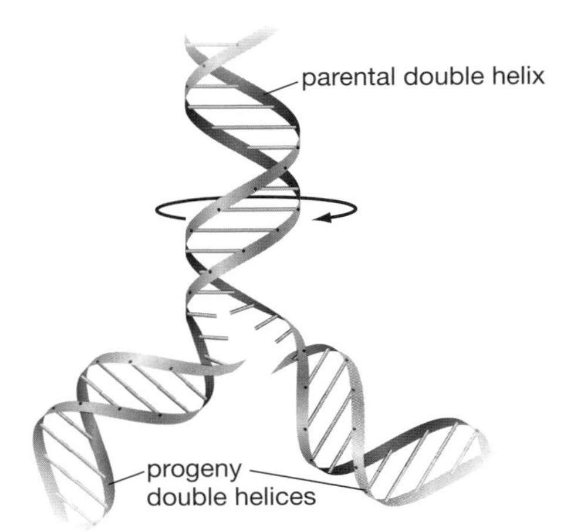

Figure 1-4. Avoidance of the intertwining of a pair of progeny DNA double helices by rotation of the unreplicated parental DNA double helix. The drawing represents a replication fork, as described in Footnote 8, proceeding in the upward direction. Intertwining of the progeny DNA molecules with one another could be avoided by rotation of the parental double helix in the direction indicated by the arrow.

the prospect of solving the untanglement problem by rotating a molecule of DNA from one end of a chromosome to the other end did not seem very plausible.

THE PROBLEM WORSENED: THE DISCOVERY OF DNA RINGS

The discovery of ring-shaped DNA molecules in the 1960s changed the nature of the untanglement problem. These molecules, often called "circular DNA" or "cyclic DNA," were initially found in virus particles.[9] Of particular significance was the 1963 finding, made in the laboratories of Renato Dulbecco and Jerome Vinograd (1913–1976), that the DNA of a polyoma virus that infects animals is a double-stranded ring with intact strands (Fig. 1-5).[10] This observation, the existence of ring-shaped DNA molecules, underscored the conceptual as well as the actual difficulty of unentangling DNA.

9. In 1962, a single-stranded DNA from the *E. coli* virus ϕX174 became the first known ring-shaped DNA, through the work of Walter Fiers and Robert L. Sinsheimer (1962. *J. Mol. Biol.* **5:** 424–434).

10. Two papers, written by researchers at two laboratories at the same institution and published in the same journal 2 mo apart, reported nearly identical studies (Dulbecco, R. and Vogt, M. 1963. *Proc. Natl. Acad. Sci.* **50:** 236–243; Weil, R. and Vinograd, J. 1963. *Proc. Natl. Acad. Sci.* 730–738). In the first of these papers, Renato Dulbecco and Marguerite Vogt stated that "This work will be supplemented by studies with the analytical ultracentrifuge," citing the Weil and Vinograd paper as being "*in preparation.*" Weil and Vinograd, in their turn, acknowledged Dulbecco for providing the polyoma virus for their study, and for providing them with a copy of the Dulbecco and Vogt manuscript before its publication.

Figure 1-5. Representation of a DNA ring in which both strands are intact (a "covalently closed DNA ring"). Because of the double-helical structure of DNA, such a ring can be viewed as containing a pair of multiply linked, single-stranded rings of complementary nucleotide sequences. Polyoma virus DNA contains ~5000 base pairs in a double-stranded ring; if this DNA assumes the classical Watson–Crick structure, the two strands revolve around each other about 500 times in the viral DNA ring.

A decade before the discovery of ring-shaped DNA, Watson and Crick had admitted in their second and third publications on the DNA double helix that the "difficulty of untwisting is a formidable one." But they did not think that uncoiling the strands of a double-helix structure would pose any fundamental problem.[11] If, however, a plectonemic double-stranded DNA is in the form of a ring (as shown in Fig. 1-5), then its two component single-stranded rings are linked and cannot possibly come apart without breaking at least one of the rings. The very conclusion that the DNA of the polyoma virus is in the form of a ring with two intact circular strands was based on the finding that the two strands of polyoma DNA did not come apart when the DNA was exposed to conditions that disrupt base pairing.[10]

In the same year, 1963, several striking images of the DNA of the bacterium *Escherichia coli* also appeared. John Cairns used a clever strategy to obtain images of very long DNA molecules embedded in an overlaid film of photographic emulsion on a coated microscope slide. Cairns did this by first incorporating tritium into *E. coli* DNA. (The nucleus of the usual hydrogen atom, ^{1}H, has just one proton, whereas the nucleus of its radioactive isotope tritium, ^{3}H, contains one proton and two neutrons.) When the tritium atoms incorporated into the DNA of *E. coli* subsequently decayed, they emitted β particles (high-speed electrons) that caused the deposition of silver grains in the photographic emulsion near the sites of these radioactive decays, thereby tracing out the shape of the DNA and yielding an "autoradiogram" of it. From such images, Cairns deduced that the entire chromosome of *E. coli*, several million base pairs long, likely existed in the form of a ring that replicates semiconservatively from a unique region of the parent DNA ring.[12]

11. Watson, J.D. and Crick, F.H.C. 1953. *Nature* **171:** 964–967; 1953. *Cold Spring Harb. Symp. Quant. Biol.* **18:** 123–131.

12. Cairns, J. 1963. *J. Mol. Biol.* **6:** 208–213.

A QUIESCENT PERIOD BEFORE THE SHIFTING TIDES

As initially conceived by Delbrück, the untanglement problem arose from the semiconservative replication of a plectonemically wound two-stranded DNA structure. Any hope that replication might not require a separation of the strands was dashed in 1958, when Matthew Meselson and Frank Stahl published their celebrated results, which showed unequivocally that *E. coli* DNA replicated semiconservatively.[13] The entanglement problem was further heightened by the 1963 work of Dulbecco and Vinograd on polyoma DNA, which had shown convincingly that the two strands in a DNA double helix must be intertwined to some extent, if not exactly to the extent required by the Watson–Crick structure. Looking back, it seems surprising that the DNA untanglement issue, so forcefully promoted by Delbrück at the birth of the concept of a plectonemic DNA double helix, was not tackled more aggressively in the following years. Several attempts, however, were made in the mid- to late 1950s. Delbrück himself had suggested one solution, which turned out to be off track. George Gamow (1904–1968), the physicist and cosmologist who posited the "Big Bang" as the source of the universe, proposed another solution for the problem, but it too was doomed.

By the 1960s, however, it seemed that Delbrück's objection to the double-helix structure had largely been forgotten, and even Delbrück himself had apparently turned his attention away from this problem.[14] Moreover, neither Dulbecco nor Vinograd, then working at the California Institute of Technology, where Delbrück was also residing, elaborated on the problem of unlinking the parental strands in a polyoma DNA ring in their 1963 publications. Similarly, although Cairns' spectacular images of ring-shaped *E. coli* DNA molecules were spreading like wildfire in the molecular biology community, there was very little discussion of the problem of how the strands of a ring-shaped chromosome became untwined during DNA replication.

Toward the end of the 1960s, however, a subtle shift occurred in the approach to the problem of DNA untanglement. Before this, the problem had mainly attracted the attention of physicists and chemists who tended to focus on a structural solution to the problem. But by the late 1960s, biochemists were beginning to consider solutions that rely on the actions of enzymes—protein catalysts known to promote a variety of reactions in the biological world.[15]

---

13. A detailed account of the experiment can be found in the late F.L. Holmes's 2001 book *Meselson, Stahl, and the replication of DNA: A history of "the most beautiful experiment in biology"* (New Haven, Yale University Press).

14. There was apparently a $5 bet between Crick and Delbrück on the plectonemic nature of the DNA structure (see Watson, J.D. 2002. *Genes, girls, and Gamow*, p. 43. Alfred A. Knopf, New York). It is not known whether Delbrück ever paid up, and if so, whether it was the data or circumstance that caused him to concede.

15. It was found in the 1980s that certain ribonucleic acids (RNAs) could also act as biological catalysts; such RNAs are termed ribozymes.

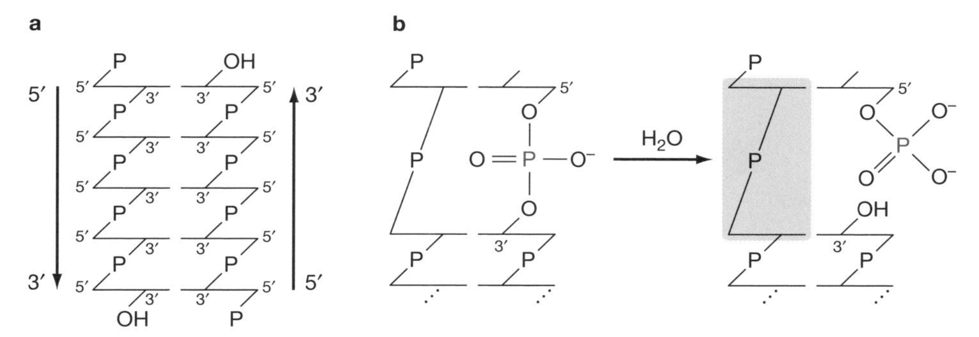

Figure 1-6. Solving the DNA untanglement problem during replication by nicking one of the two DNA strands. (*a*) A simplified representation of a double-stranded DNA. Each zigzag line represents a DNA backbone strand consisting of phosphate moieties (P) linked to the 5′ carbon atoms of ribose moieties, whose 3′ carbon atoms are linked to other phosphate moieties (·····P-5′– 3′- P-5′– 3′- P·····) (compare with Fig. 1-1b). The horizontal lines connecting the two zigzag lines represent the base pairs joined by hydrogen (H) bonds at the gaps between the horizontal lines. (*b*) Expanded section of the DNA representation in *a*, to show the cleavage of the bond linking one phosphoryl moiety to the 3′ carbon of an adjacent ribose group. DNase catalyzes the attack of the phosphorous atom by the oxygen atom of a water molecule (H_2O), to break a P–O bond in the DNA backbone, yielding either a 5′- or a 3′-phosphate group that carries a double-negative electronic charge ($-PO_4^{=}$), depending on the particular DNase, and a 3′- or 5′-OH group (in the figure, a 5′-phosphate and a 3′-OH group are produced). The shaded area opposite the nick is shown in greater detail in Fig. 1-7.

One class of enzymes that stood out as attractive candidates in solving the DNA untanglement problem was the so-called deoxyribonucleases, or DNases. These enzymes, which were known long before the double-helix structure was proposed, catalyze the hydrolysis of the chemical bonds that link the components in the backbone strands of a DNA molecule. In this hydrolysis, a water molecule attacks a phosphorous, breaking a bond in the DNA backbone between the phosphorous and the sugar moiety to which it is attached (Fig. 1-6). There are two major categories of these enzymes: Those that "nibble" from the ends of DNA strands are called DNA exonucleases, and those that introduce breaks within DNA strands are called DNA endonucleases. Among the endonucleases, those that can break a single strand of the DNA double helix, rather than breaking both strands at once, appeared to be the most attractive candidates for resolving the DNA untanglement problem. In a replicating double-stranded DNA, a few strategically placed breaks ("nicks") in one of the strands could in principle solve the problem of how to separate the interwound strands. Chemists had long known that there is very little resistance to the rotation of two parts of a molecule separated by a single bond, like the bond between the two CH_3 groups in ethane, $H_3C—CH_3$. Accordingly, if a nick is made on either side of a DNA segment within a long DNA molecule, then each of several single bonds on the strand opposite each nick can serve as a swivel for rotation of the DNA segment

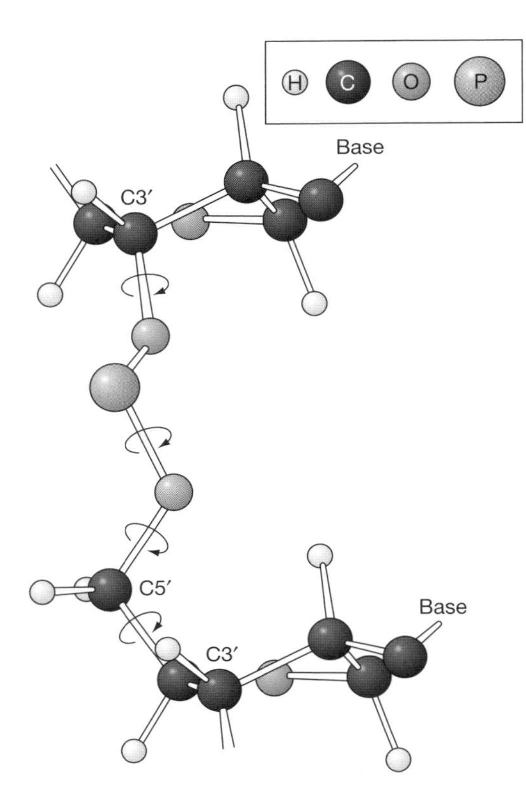

Figure 1-7. A structural model depicting the backbone bonds that link two neighboring bases in the strand directly across from the nick in its complementary strand (the shaded area in Fig. 1-6b). The carbon atoms are represented by black spheres, the hydrogen atoms by white spheres, the oxygen atoms by gray spheres, and the phosphorus atoms by blue spheres. The sugar rings are shown in the puckered conformation in which they exist in a typical DNA molecule (see legend to Fig. 1-1b). The curved arrows are placed around single bonds in the DNA backbone that connect two adjacent bases in the DNA strand. Rotation around these single bonds can take place in either direction and does not necessarily have to occur in the direction indicated by the arrows.

relative to the rest of the molecule (Fig. 1-7). In other words, in the presence of a few nicks, the untwining of one part of a long DNA could proceed without rotating the entire DNA.

When in 1968 Ju-ichi Tomizawa and Tomoko Ogawa attempted to find such strand breaks in a replicating bacteriophage λ DNA, they found that fewer than one, and probably none, could be detected in either of the two strands of this DNA, each of which comprised 50,000 nucleotides.[16] Nevertheless, their findings could not rule out the possibility that breaks in DNA strands are indeed the means that solve the DNA untanglement problem. The breaks, or nicks, might be of a mobile type—that is, they might be created at one site, disappear upon rejoining of the ends of the nicked or broken DNA strand by an enzyme called DNA ligase, and occur again somewhere else in the DNA molecule ahead of a replication fork (Fig. 1-4). It would indeed be a daunting task to prove or disprove the presence of such mobile replication swivels.

16. Tomizawa, J.-I. and Ogawa, T. 1968. *Cold Spring Harb. Symp. Quant. Biol.* **33:** 533–551.

"Fortunately, science" wrote the English chemist Humphrey Davy (1778–1829) "...is neither limited by time nor by space." By 1968, when Tomizawa and Ogawa showed that very few if any strand breaks were present in the DNA of a replicating phage λ, hints about Nature's way of solving the DNA untanglement problem were about to emerge. It would soon be revealed that the idea of mobile replication swivels was close to the mark, but the particular kinds of mobile swivels found would be entirely unexpected.

CHAPTER 2

Nature's Solutions
The DNA Topoisomerases

"If a little fact will not fit in, they throw it aside. But it's always the facts that will not fit in that are significant."

Hercule Poirot, in Agatha Christie, *Death on the Nile*

A POPULAR TRICK IN THE REPERTOIRE OF MANY amateur and professional magicians is the "linking ring" trick: The magician first lets a spectator examine a few solid rings, then does some hocus pocus, and, lo and behold, two rings are suddenly linked! Inspection of the linked rings by the spectators reveals no hidden contraption in either ring for passing one through the other. A bit more prestidigitation, and the linked rings again come apart. As the performance progresses, there ensues a dazzling display of the linking and unlinking of many rings.

The popularity of this trick lies in its simplicity. There is no need for the magician to explain the obvious: that separate rings cannot be linked and linked rings cannot come apart; hence the magician's ability to do the impossible invariably impresses the audience. What the magician does not disclose is that the separate and linked rings, both of which have been painstakingly inspected by the spectators, are never the same. In truth the separate rings never become linked and the linked ones never come apart; the trick instead relies on a simple premise of magic: that the hands can be faster than the eyes.

By contrast, the replication of any ring-shaped DNA molecule poses a truly formidable problem: Semiconservative replication of such a DNA ring requires that its two strands, each entwined about the other numerous times, come cleanly apart. Unlike the magician, Nature must provide an honest solution during each replication cycle. How is this accomplished?

The discovery of DNA rings projected the DNA untanglement problem into the realm of topology, a branch of mathematics that deals with properties of geometric structures that are preserved through deformations that do not involve the structures being torn or broken apart. Linkage between two rings is such a property: One or both members of the pair of linked rings (also called a "catenane") can be stretched, shrunk, or twisted out of shape, but such deformations do not alter the link between them. That link is termed a "topological invariant" in the jargon of mathematics. Whereas subjects of topology often seem exotic and unfathomable to all but the hardcore enthusiasts (topologists), the basic topological principles relevant to the DNA untanglement problem can be easily grasped. Although this may require some patience, it can prove well worthwhile. Nature's solutions to seemingly complex problems are often simple and elegant, and rarely is this elegance more evident than in the solutions provided for the DNA untanglement problem.

A PRELUDE: THE TOPOLOGY OF RINGS AND THE DISCOVERY OF SUPERCOILED DNA

In 1963, the discovery of naturally occurring DNA rings composed of two intertwined single strands (see Chapter 1) was about to provide a powerful tool for studying the DNA untanglement problem. In that year, investigators in the laboratories of Renato Dulbecco and of Jerome Vinograd observed that DNA purified from polyoma virus contained two components, termed "Component I" and "Component II," that were readily distinguishable by sedimentation in an ultracentrifuge. Component I, the major and faster sedimenting component, was believed to be the ring form that we described in Chapter 1; Component II, the minor and slower sedimenting component, was thought to be a linear form of the same DNA.[1] Roger Weil and Vinograd observed yet a third component, Component III, which was present in small amounts in various preparations and sedimented even more slowly than Component II. The nature of this third species was at the time left unexplained.

Appended to the 1963 paper by Weil and Vinograd was a note by Walter Stoeckenius describing the appearance of polyoma DNA molecules when viewed in an electron microscope. Stoeckenius observed both ring and linear forms in ratio close to that expected from the sedimentation analysis. However, Component III, which was presumed to have a different size and/or shape than Components I or II, was not apparent in the electron micrographs. Among the ring-shaped DNA molecules Stoeckenius also noticed a high proportion of "tightly coiled" forms, having what appeared to be cross-links between different points along their contours (Fig. 2-1). Because this kind of tight coiling was rarely seen in the linear DNA molecules in these electron micrographs, Stoeckenius thought that the ring-shaped DNA molecules were somehow more prone to cross-linking by proteins.

1. Vogt, M. and Dulbecco, R. 1963. *Proc. Natl. Acad. Sci.* **49:** 171–179; Weil, R. and Vinograd, J. 1963. *Proc. Natl. Acad. Sci.* **50:** 730–738. See Footnote 10 in Chapter 1.

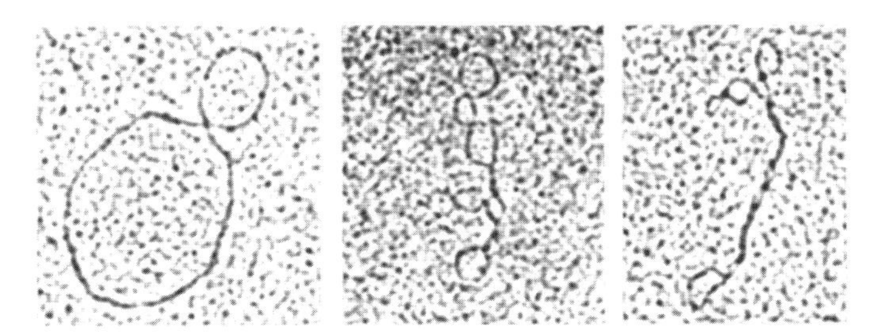

Figure 2-1. Selected images of polyoma viral DNA by electron microscopy. The majority of the DNA molecules were in the form of rings; the *leftmost* ring represents the "extended cyclic form," which showed very few self-crossings, and the two to its right the "tightly coiled cyclic form." A minor population of linear DNA molecules with clearly discernible ends were also seen, but none is depicted here. See the text for details. Each polyoma virus DNA contains ~5000 bp and has a contour length of ~1.7 μm. Before viewing, the DNA molecules were first coated with a protein, and then a very thin metal film was deposited onto the coated molecules by vaporizing the metal in a high vacuum. These treatments enhanced the visibility of the DNA molecules in the electron microscope. See Stoeckenius, W. 1963. *Proc. Natl. Acad. Sci.* **50:** 737–738. (Reprinted, with permission, from Weil, R. and Vinograd, J. 1963. *Proc. Natl. Acad. Sci.* **50:** 730–738.)

Vinograd decided to dig a bit deeper into the properties of these three forms of polyoma DNA. Roger Radloff, a graduate student in Vinograd's laboratory, painstakingly separated Components I and II by using a preparative ultracentrifuge that fractionates large molecules that sediment differently. Philip Laipis, an undergraduate, was asked to examine these purified components in an electron microscope. Laipis made a very puzzling finding: DNA molecules in the fraction containing Component I were indeed in the form of rings, as expected, but DNA molecules in the fraction containing Component II, the putative linear form, also appeared ring-shaped! Interestingly, there was a significant difference in the appearance of the rings in these two fractions: Component I consisted mostly of the "tightly coiled" form described by Stoeckenius, whereas Component II was seen mostly as rings with a more open appearance and very few intramolecular cross-links. Vinograd soon deduced the likely structural basis for this tight coiling of Component I molecules.[2]

Recall from our discussion in Chapter 1 that a double-stranded DNA ring, like that of polyoma DNA, is expected to contain two linked, single-stranded DNA rings

2. This seminal discovery was reported in Vinograd, J., et al. 1965. *Proc. Natl. Acad. Sci.* **53:** 1104–1111. The authors of the paper included almost all members of Vinograd's small research group then at Caltech: Lebowitz was a postdoctoral fellow, Radloff a graduate student, Watson a technician, and Laipis an undergraduate. This work also launched Vinograd, then a research associate in his 50s, to international fame. However, readers knowledgeable in the topology of DNA rings might be able to spot two errors in this celebrated paper (incorrect handedness of the supercoils and the wrong quantitative relation between the number of crossovers removed and the DNA base pair disrupted).

(see Fig. 1-5). The order or degree of linkage between the two interlocked rings (a catenane), whether they consist of DNA strands, stainless steel cables, or rubber tubing, can be specified mathematically as a quantity called the *linking number* (*Lk*). Figure 2-2 illustrates several simple cases with $Lk = 1$, 2, and 3 between a pair of rings. Vinograd recognized that in a ring-shaped DNA double helix, *Lk* should be closely related to the helical periodicity of the double helix (how tight the helical turns are). In the Watson–Crick structure, the DNA helix makes a right-handed turn every 10 base pairs (bp); that is, it has a helical periodicity of 10 bp per turn. Thus if a DNA ring contains 4500 bp, then the two strands would be expected to coil around each other 450 times (4500 bp/10 bp). Accordingly, the expected *Lk* for the two component rings of a polyoma DNA should be ~450.

We can also follow the same reasoning in reverse. Suppose that under a particular set of experimental conditions, the most stable structure of the DNA double helix

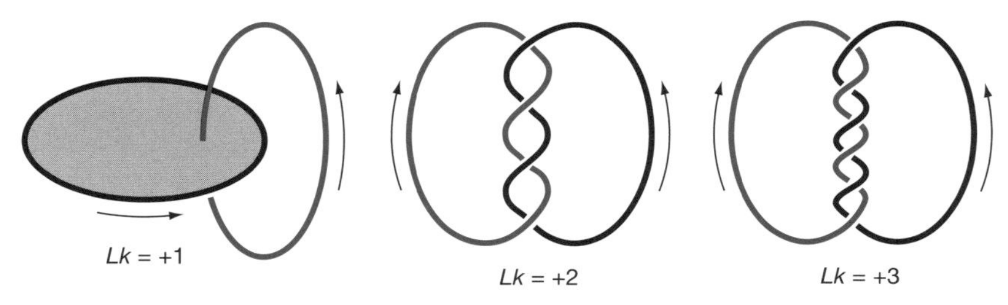

Figure 2-2. Sketches of three pairs of rings with linking numbers 1 (*left*), 2 (*middle*), and 3 (*right*). Each pair of thick lines represents two linked closed curves (rings), and the arrow marking each curve as well as the small arrow below the black ring indicates the direction or polarity of the curve (e.g., the 5'→3' direction in the case of a DNA strand). In the two *rightmost* drawings, it can be seen that the two curves wrap around each other in a right-handed way and run antiparallel to each other. The linking number of each pair of closed curves corresponds to the number of times one curve winds about the other. The case with $Lk = +1$ illustrates a more rigorous definition of *Lk* (see Pohl, W.F. and Roberts, G.W. 1978. *J. Math. Biol.* **6:** 383–402). The shaded area represents a surface bounded by the black closed curve and is called the "spanning surface" of that curve. Because the curve encircling this surface has a polarity, the two sides of this surface are different. According to the right-hand convention, when the fingers of the right hand curl in the direction specified by the polarity of the curve, then the thumb of the right hand would specify the minus to plus side of the spanning surface. In the *left* drawing, the bottom side of the shaded spanning surface defined by the black circle is the minus side and the upper side of the same surface is the positive side. The linking number *Lk* between a pair of closed curves is the net number of times one curve passes from the + side to the – side of the spanning surface of the other closed curve. In the *left* drawing, the blue line crosses once from the + side to the – side of the spanning surface of the black line, and hence $Lk = +1$. Thus *Lk* is actually an algebraic quantity and can have positive or negative values. Note that if the polarity of either ring in the *left* drawing is reversed, then *Lk* becomes –1, but that reversing the polarity of both rings does not alter the sign of *Lk*. In all cases the sign convention has been chosen in such a way that *Lk* between two antiparallel strands in a right-handed double helix DNA ring is positive.

is precisely 10.0 bp per turn, as proposed by Watson and Crick.[3] Then if a 4500-bp DNA ring has an *Lk* of exactly 450, the entire ring can happily assume the Watson–Crick helical structure with its helical axis lying in a plane. In such a case, the entire DNA ring can be laid down on a flat surface, such as the thin film the DNA molecules were placed on for examination in an electron microscope.

But what if the same DNA happens to have an *Lk* significantly different from 450, say 420 or 480? Like a torsionally unbalanced rope, such a DNA ring would coil up in space; its helical axis would no longer lie in a geometric (two-dimensional) plane, but instead would writhe in space to trace out a closed curve that often crosses over itself in its planar projections (Fig. 2-3). Vinograd recognized that if, for whatever reason, the *Lk* of a DNA ring within a cell is different from that of the same ring in its most stable helical structure outside the cell, then, upon isolation from the cell, the DNA ring would have the "tightly coiled" appearance first seen by Stoeckenius.[2]

Vinograd and his coworkers soon deduced that a DNA ring from a natural source would typically display a value of *Lk* several percent lower than that of the same DNA ring in its most stable structure under the typical experimental conditions used in laboratories.[4] The latter quantity, termed the linking number of a *relaxed* DNA ring, is often assigned the symbol Lk^0. If a break is first introduced into one strand of a double-stranded DNA ring to relieve the invariancy of *Lk*, then the nearby single bonds in the DNA strand opposite the nick could serve as a swivel (see Chapter 1, Figs. 1-6 and 1-7). Upon resealing of the break, the DNA ring would have an *Lk* corresponding to the value of Lk^0 under the resealing conditions. Thus Lk^0 serves as a reference value for a particular DNA ring under a particular set of conditions. A DNA ring with a linking number *Lk* is said to be "underwound" or "negatively supercoiled" if $Lk < Lk^0$, or $(Lk - Lk^0) < 0$, and "overwound" or "positively supercoiled" if $Lk > Lk^0$, or $(Lk - Lk^0) > 0$.[5,6] Polyoma

3. For a typical DNA, the helical periodicity in solution in a dilute aqueous solution around room temperature is actually about 10.5 rather than 10.0 (Wang, J.C. 1979. *Proc. Natl. Acad. Sci.* **76:** 200–203; Rhodes, D. and Klug, A. 1980. *Nature* **286:** 573–578). The precise value of the DNA helical periodicity does not affect the conclusions in most of the cases, and therefore for simplicity the value 10.0 will often be used in this volume.

4. Bauer, W. and Vinograd, J. 1968. *J. Mol. Biol.* **33:** 141–171; Vinograd, J., et al. 1968. *J. Mol. Biol.* **33:** 173–197.

5. The notations used here differ from those used in the original papers from Vinograd's laboratory, where the extents of supercoiling were described in terms of a parameter τ that Vinograd termed the "superhelix winding number" or the "number of superhelical turns (supercoils)." The meaning of τ was not clearly specified. From the way in which Vinograd's laboratory measured τ, however, it is clear that τ is identical to $(Lk - Lk^0)$. It should also be noted that *Lk* of a DNA ring is meaningful only if both strands are intact, and if it is an integer for a particular DNA ring; Lk^0, however, is not a topological invariant, and its value for a particular DNA ring depends on the particular experimental conditions. Also, Lk^0 is the linking number of a hypothetical or idealized state and is not necessarily an integer. For example, if the helical periodicity of a 205-bp DNA ring under a particular set of experimental conditions is precisely 10.0, then Lk^0 is 205/10 or 20.5; no real covalently closed DNA ring can have such a non-integer linking number, however, and Lk^0 can be viewed as the average of a population of completely relaxed 205-bp DNA rings with integral linking numbers.

6. Before 1974, the values of $(Lk - Lk^0)$ reported in the literature were off on the low side by about a factor of 2. The source of this large error is briefly described in Footnote 4 of the next chapter (for details, see Wang, J.C. 1974. *J. Mol. Biol.* **89:** 783–801).

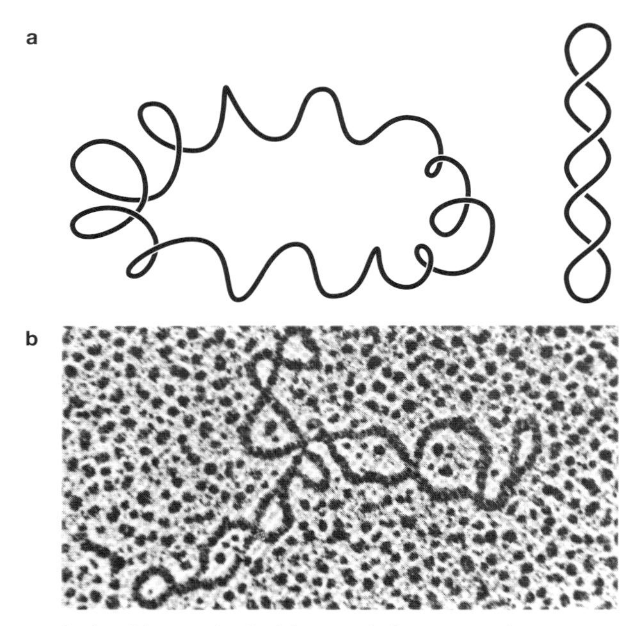

Figure 2-3. (*a*) Two idealized forms of a double-stranded DNA ring, here represented by a thick line, when the linking number of the ring deviates from that of the same ring in a relaxed state. In the *left* drawing, the line can be visualized as wrapping around a donut surface, representing what is termed "solenoidal supercoiling" or "toroidal supercoiling"; in the *right* drawing, the line coils upon itself (is "interwound"), and is termed "plectonemic supercoiling." Both representations are for a negatively supercoiled DNA; the two representations are topologically equivalent despite their clear geometric differences. It is interesting to note that in the solenoidal representation, the negatively supercoiled DNA follows a *left-handed* helical path around the imaginary donut surface, whereas in the interwound or plectonemic configuration the lines going up and down are wrapped in a *right-handed* way about each other. (Redrawn, with permission, from Cozzarelli, N. R., et al. 1990. In *DNA topology and its biological effects* [ed. N.R. Cozzarelli and J.C. Wang], pp. 139–184. Cold Spring Harbor Laboratory Press, Cold Spring Harbor, New York.) See also Bauer, W. and Vinograd, J. 1968. *J. Mol. Biol.* **33:** 141–171. (*b*) Electron micrograph of a negatively supercoiled plasmid DNA containing about 10,000 bp. (Reprinted, with permission, from Wang, J.C. 1982. *Sci. Am.* **247:** 94–109.)

and other DNA rings from natural sources were thus defined to be negatively supercoiled by Vinograd and his associates.

For a ring-shaped DNA with intact strands, the helical structure of DNA favors a particular linking number Lk^0 under a particular set of experimental conditions. As we discussed, if the most stable DNA structure under these conditions is one having 10.0 bp per helical turn, then Lk^0 is expected to be equal to one-tenth the number of base pairs in the ring. Consider a DNA ring with a linking number Lk different from Lk^0. This molecule cannot assume the most stable DNA structure, because Lk, being a topological invariant, cannot be changed to Lk^0 without the transient introduction of a break into at least one of the two DNA strands. Thus the spatial coiling (or writhe) of a DNA ring for which *Lk* deviates significantly from Lk^0 represents the best compromise between the tendency of a DNA double helix to assume its preferred helical geometry and the topological requirement of an invariant linking number.

With the realization that Component I of polyoma virus DNA is the negatively supercoiled form, all experimental data described earlier could be understood. Breaking one strand of Component I would remove the topological constraint and yield the less compact Component II, which sediments more slowly than Component I. Component III turned out to be the linear form seen in Stoeckenius' electron micrographs; but the preconceived notion that Component II was the linear form had led to the conclusion that "Component III could not be identified in the electron micrographs." The old saying "to see is to believe" should be practiced with caution: What one sees and what one thinks one sees are not always the same thing!

A SURPRISING FINDING AND A PLAUSIBLE EXPLANATION QUICKLY DASHED

An important clue to Nature's way of dealing with the DNA untanglement problem came in the late 1960s. The discovery in 1965 of supercoiled polyoma DNA, with subsequent studies of other DNA rings purified from various cells, had raised an inevitable question: Why were all of these DNA rings negatively supercoiled? Although Vinograd was able to attribute the negative supercoiling of a DNA ring to its reduced linking number relative to Lk^0 (the linking number of the same DNA ring in its relaxed state), the question why this should be so was left hanging.

In an attempt to gain some insight into the cause of negative supercoiling of DNA rings from natural sources, I conducted a number of experiments in around 1968 to measure the quantity $(Lk - Lk^0)$ for a family of DNA rings of different sizes from *Escherichia coli* cells, to see how $(Lk - Lk^0)$ would vary with the size of a DNA ring or with changes in the growth condition of the cells.[7] I observed that the values of $(Lk - Lk^0)$ were generally proportional to the sizes of the DNA rings, with minor alterations but no dramatic changes in $(Lk - Lk^0)$ per unit size of the rings. There was, however, one "little fact" that did not fit the general pattern: Whereas typical samples of intracellular phage λ DNA rings contained only a small fraction of relaxed DNA rings,[7] one particular preparation of the DNA rings was found to contain mostly relaxed rather than negatively supercoiled DNA.[8]

This last finding was very surprising. Not only did it not fit the general pattern exhibited by the $(Lk - Lk^0)$ values, but all previous preparations of the same DNA, as

7. Wang, J.C. 1969. *J. Mol. Biol.* **43:** 263–272. With hindsight, the main results of that paper were not as significant as the unexpected finding described in the text.

8. A historic account has been provided in Wang, J.C. 2004. Reflections on an accidental discovery, In *DNA topoisomerases in cancer therapy: Present and future* (ed. T. Andoh). Kluwer/Plenum, New York.

well as subsequent repetitions of the same preparation, yielded samples that contained mostly negatively supercoiled rather than relaxed DNA rings. So what had happened in that particular instance? Upon thorough review of the experimental records, a likely cause of this singular exception was revealed.

In our standard protocol for such preparations, small ring-shaped DNA was isolated from *E. coli* cells by ultracenetrifugation of the cell lysate. (Even though a phage λ DNA ring contains ~50,000 bp, it is considered a "small" ring when compared with the much larger *E. coli* chromosome, containing more than 4 million bp.) *E. coli* cells in a certain volume of culture were first briefly spun in an ultracentrifuge to pellet the cells. The cell pellet was resuspended in a smaller volume of an appropriate buffer, and a gentle detergent was added to the cell suspension to open the cell membranes, thus allowing the λ DNA rings to leak out of the cells. The resulting cell lysate was then spun again in the ultracentrifuge, this time at a much higher speed, to pellet the cell debris and the much larger *E. coli* chromosomal DNA. This would leave most of the λ DNA rings in the resulting supernatant, ready for subsequent purification.[7]

The steps described above were usually performed at ~0°C, a temperature often favored by biochemists in the preparation of biological materials, in order to minimize chemical changes or the likely actions of enzymes such as DNases that break DNA. In the particular preparation that yielded mostly relaxed λ DNA rings, however, the normal procedure was not followed; the high-speed centrifugation step was accidentally performed at 20°C rather than 0°C, and for a much longer period (2.5 h rather than the usual 20 min).[8]

Could it be then that during the longer period at 20°C, a low level of a particular DNase activity in the cell extract might have slowly introduced nicks into the double-stranded DNA rings? Such nicks would allow the DNA to swivel around its helical axis, and relaxed DNA rings would be produced upon their resealing by DNA ligase, an enzyme that had been discovered a year earlier (as discussed in Chapter 1). Indeed, a quick check showed that the kind of *E. coli* cell extracts used in the DNA ring preparations contained ample DNA ligase activity.[7] Whereas the opposing actions of a DNA endonuclease that introduces nicks and an excess of DNA ligase that rapidly seals all such nicks would leave no apparent change in a linear DNA, in a supercoiled DNA ring some or all of the negative supercoils might escape after one or more cycles of nicking and resealing by a DNA endonuclease and DNA ligase, respectively.

A failed attempt to purify this putative endonuclease activity soon revealed, however, that my idea of competing DNase–DNA ligase activities was completely off the mark. The cell extracts contained plenty of DNA ligase activity, but the activity could be readily separated from the supercoil relaxation activity by ion-exchange chromatography, in which proteins bound to an ion-exchange resin were gradually washed off the resin with solutions of increasing salt concentration. In this approach, the supercoil relaxation activity and DNA ligase were washed off the resin at different salt concentrations, and their separation showed that the component in the cell extract that removed DNA negative supercoils had nothing to do with the presence of DNA ligase in the cell extract. If DNA ligase was not involved, what might account for the disappearance of the supercoils upon treatment of the phage λ DNA rings with the cell extract?

"Eliminate All Other Factors, and the One Which Remains Must Be the Truth"

There were several other plausible explanations for the apparent disappearance of the negative supercoils. First, one or more components in the cell extract could have changed the helical structure of the input DNA so as to reduce the pitch of the double helix. The DNA double helix was known to exist in several forms with different helical geometries. The classical 10 bp-per-right-handed-turn structure described by Watson and Crick became known as the DNA B-helix structure; among the others, the A-form double helix is less tightly wound and has a periodicity of 11 bp per helical turn. Could exposing a DNA molecule to a certain protein in the *E. coli* extract, for example, change a portion of the DNA from the normal B structure to the A structure? Consider as an example a DNA ring of 11,000 bp. In the B-helix geometry, the DNA ring would be expected to have an Lk^0 of 11,000/10 or 1100. This DNA ring with an Lk of 1050 would assume a negatively supercoiled form. But if half of the base pairs of an 11,000-bp DNA ring were in the B form and the other half were in the A form, the expected Lk^0 would be (5500/10) + (5500/11) or 550 + 500 = 1050. In other words, an 11,000-bp DNA ring with Lk = 1050 would be in a completely relaxed state if it had such a mixed helical geometry.

Studies of base pair geometries, however, dismissed this idea. DNA base pairs having different helical geometries were known to differ in their absorption of polarized ultraviolet (UV) light. Circular dichroism (CD) spectroscopic analysis, which can distinguish these differences, revealed no detectable variation in DNA samples before and after their treatment with the partially purified activity from *E. coli* cell extracts. Both treated and untreated DNA samples showed the same CD spectrum characteristic of the DNA B-helix.

If there was no change in the helical geometry of a significant part of the DNA ring, could a much shorter stretch of the DNA have somehow become completely unpaired upon treatment of the DNA with the cell extract? If 500 bp of the 11,000-bp ring were unpaired, then Lk^0 of the DNA would be expected to be (11,000 – 500)/10 = 1050. Such a DNA, with an Lk = 1050, would behave like a relaxed DNA ring in terms of its spatial coiling and its sedimentation velocity in an ultracentrifuge. An unpaired region of 500 bp in an 11,000-bp DNA ring could easily escape detection by electron microscopy, especially if the unpaired nucleotides were distributed in several small bubbles rather than concentrated in one contiguous region.

The presence of short single-stranded regions in a DNA ring would be detectable, however, by the use of DNA endonucleases that specifically break the backbone bonds in single-stranded DNA. When I exposed samples of double-stranded DNA rings to an endonuclease before and after their treatment with the partially purified activity from cell extracts, the treated DNA was not found to be especially vulnerable.

Apparently, then, the treatment of a supercoiled phage λ DNA with the partially purified activity from cell extracts did not add much single-stranded character to it.

If the loss of negative supercoils was due to a change in the helical geometry of the DNA resulting from its binding to a protein, or from a chemical modification, such as DNA methylation (the addition of methyl [CH_3] groups to DNA bases), then the density of the DNA might also change. A very sensitive method for detecting density changes in DNA, unveiled in the Meselson–Stahl experiment mentioned in Chapter 1, was widely in use in 1970. But here, too, the use of density gradient centrifugation revealed no density difference between treated and untreated DNA samples.

The results of these experiments thus rejected what might be considered, at that time, the most likely interpretations for the apparent loss of negative supercoils. And as purification of the supercoil-removal activity proceeded, these possibilities became even less attractive. I found that the partially purified material responsible for supercoil removal was protein-like, and so potent that tiny amounts of it were sufficient to remove nearly all of the negative supercoils in a λ DNA ring of 50,000 bp. It was difficult to imagine how a small amount of a protein could cause a change in the helical structure of a large stretch of DNA through base modification or could otherwise disrupt hundreds of base pairs in each DNA ring. It was therefore perhaps time to apply the dictum of the legendary detective Sherlock Holmes: "Eliminate all other factors, and the one which remains must be the truth".[9] What possibility remained?

TRANSESTERIFICATION: SWAPPING ONE ESTER BOND FOR ANOTHER

The experiments summarized above indicated that exposure to the protein fraction purified from *E. coli* cell extracts did not significantly change the helical structure, nor form single-stranded regions, in supercoiled phage λ DNA. Yet the supercoiled DNA was converted to the seemingly relaxed form. Further, this conversion did not involve the combined actions of DNA endonuclease and ligase. So how could the negative supercoils simply vanish?

The one remaining explanation was that the *E. coli* activity, by then given the obscure name "ω protein," might catalyze the *transient* breakage of a DNA strand.[10]

---

9. Sherlock Holmes, in Doyle, A.C. 1890. *The Sign of Four*.

10. Wang, J.C. 1971. *J. Mol. Biol.* **55:** 523–533. There had been occasional speculation about why we chose this particular Greek character. The protein was called "ω" because the relaxation of negative supercoiled DNA was initially followed in an analytical ultracentrifuge by monitoring the change in the sedimentation velocity of a supercoiled DNA ring, and a key parameter in centrifugation is the angular velocity, ω. We performed more than a thousand centrifugation runs during the first purification of this activity. The designation "ω" also provided an intellectual link between the study of this protein and the DNA untanglement problem, often associated with the problem of rapidly rotating the DNA ahead of a replication fork at an angular velocity that is comparable to that of an ultracentrifuge.

If incubation with the *E. coli* ω protein had left no detectable interruption in either strand of the double-stranded DNA ring, then any break must have been rejoined by a second activity distinct from *E. coli* DNA ligase. In principle, this second activity could reside on a separate protein; but it seemed more likely to reside on the same protein, because the two activities did not seem separable during purification.

How might an enzyme do both jobs? Could a single protein, the ω protein, have both DNase and ligase activities? I rejected such a possibility, however, based on the known requirements of the *E. coli* ω protein: It required only magnesium ions for its removal of DNA negative supercoils. To understand why such a spartan requirement undermined the likelihood of the DNase–ligase idea, it is important to grasp a fundamental difference between spontaneous and nonspontaneous reactions in nature.

Spontaneous and nonspontaneous reactions are very common in everyday life. A waterfall rushing down a cliff is a spontaneous process, and its energy can be harnessed by passing the flowing water through a generator or a water mill. By contrast, water flowing from a deep well into a storage tank on the roof of a tall building is a nonspontaneous process; here electric energy, or human labor, is required to pump the water up. In general, a spontaneous process is one from which energy can be harnessed to do useful work—it is energetically favorable; a nonspontaneous process is energetically unfavorable and requires the input of energy to make it happen.[11] Hydrolysis of a DNA backbone bond, in which a water molecule attacks a phosphodiester bond in a DNA strand and leaves a pair of phosphoryl and hydroxyl groups at the broken DNA ends (see Fig. 1-6), can occur spontaneously. Indeed, hydrolysis will eventually break down a DNA molecule completely, albeit extremely slowly. The rate of the hydrolysis can be increased at higher temperatures or in the presence of chemicals like heavy-metal ions, and a DNase can serve as a powerful catalyst that greatly accelerates this process.

By contrast, the reverse reaction, the formation of a DNA phosphodiester bond from an adjacent phosphoryl and hydroxyl group, is nonspontaneous. The expulsion of a water molecule from a phosphoryl and hydroxyl pair at a strand break is an energetically unfavorable task in an aqueous solution in which water is everywhere. So how does DNA ligase accomplish this task of sealing a nick in DNA? The enzyme couples this reaction to a spontaneous process—the breakdown of ATP (adenosine triphosphate) or NAD^+ (nicotinamide adenosine diphosphate); the cofactor involved depends on the particular DNA ligase.[12] Breaking down ATP or NAD^+ releases its

11. The relation between the spontaneity of processes and energy flows is the subject of a branch of science known as *thermodynamics*. The term *energy* is used here in a loose sense, and connoisseurs would prefer the more precise term *free energy*, which takes into account both the natural tendency of going from a higher energy state to a lower one (water flowing down a hill) and from a more ordered state to a less ordered one (a drop of ink spreading in a pail of unstirred water).

12. DNA ligases of bacteria typically use NAD^+ as the cofactor, and other DNA ligases, including those encoded by bacterial viruses, typically use ATP. In all cases the first step is to charge up the enzyme by forming an enzyme–adenylate (enzyme–AMP) intermediate. In this covalent intermediate, an AMP, from either NAD^+ or ATP, is linked to a lysine amino group through a phosphoamide bond.

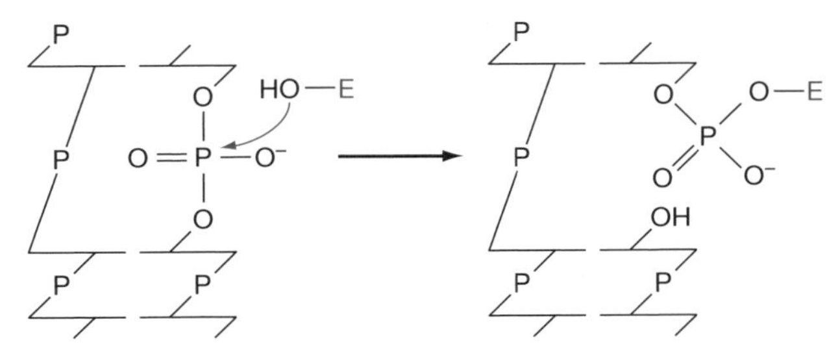

Figure 2-4. A bond-swapping mechanism postulated in 1971 for the breakage of a DNA strand by the *E. coli* ω protein. See Fig. 1-6b for explanation of the representation used for the double-stranded DNA segment. One phosphoryl group in the backbone of the strand on the right is depicted. E–OH represents an *E. coli* ω protein, and –OH indicates the hydroxyl group of a particular amino acid residue of the polypeptide. In the postulated mechanism, the oxygen atom of the hydroxyl group attacks the phosphorus atom (P) in the phosphoryl group to break an O–P bond in the strand and simultaneously form an O–P bond between the DNA and the ω protein. See Wang, J.C. 1971. *J. Mol. Biol.* **55:** 523–533.

stored chemical energy, which is then used to promote the otherwise nonspontaneous ligation of the broken DNA strand.[13]

The removal of negative supercoils by the *E. coli* ω protein, however, requires no ATP or NAD^+, nor any other compound of that category. It therefore seemed highly unlikely that the mechanism involved hydrolysis (if that were the case, then resealing of the broken DNA strand would surely require a cofactor). Whatever the true mechanism, this reasoning led me, in 1971, to a propose that the *E. coli* ω protein performs its magic by the scheme illustrated in Figure 2-4: bond swapping. In this scheme, the breakage of a DNA backbone bond is accompanied by the formation of a DNA–protein covalent bond, and the re-formation of the DNA backbone bond is accompanied by the breakage of the DNA–protein bond.[10] Specifically, I postulated that an oxygen atom (O) of the hydroxyl (OH) group of a particular amino acid side-chain of the enzyme would first attack a DNA phosphorus (P), joining itself to the DNA phosphorus and at the same time breaking a P—O bond in the

13. A DNA ligase-mediated joining of a DNA phosphoryl and a DNA hydroxyl group is sometimes described as the coupling of a nonspontaneous reaction, the elimination of a water molecule from the pair, and a spontaneous reaction consisting of the hydrolysis of ATP to AMP and pyrophosphate (PPi) or NAD^+ to AMP and nicotinamide (NMN^+). This language is formally correct, but it is mechanistically misleading. In the actual reaction, the enzyme first passes the AMP of the enzyme–AMP complex (see Footnote 12) to the phosphoryl group at one DNA end. The oxygen atom of the hydroxyl group at the other DNA end is then placed in a position to attack the phosphorus atom of the AMP-linked DNA 5′-phosphoryl group, forming a DNA phosphodiester bond and releasing the AMP. The important point is that DNA ligase keeps water out of the various reaction steps. If a water molecule managed to sneak into the catalytic pocket of the enzyme and attacks the AMP-linked phosphoryl group, hydrolysis could occur and the reaction would then simply release the AMP without forming the DNA phosphodiester bond.

DNA backbone. The protein would now be covalently linked to a DNA phosphorus at one end of the broken DNA strand, leaving at the other end of the broken DNA strand a hydroxyl group on a deoxyribose. In the subsequent rejoining step, the oxygen of the hydroxyl group on the pentose sugar (deoxyribose) would attack the DNA phosphorus covalently attached to the protein, thereby breaking that bond and re-forming the DNA backbone bond. In both the DNA breaking and rejoining reactions, the breakage of what chemists call an ester bond (here a P—O bond) is accompanied by the formation of another ester bond (also a P—O bond in the scheme shown in Fig. 2-4). This bond-swapping reaction, in which one ester bond is broken while another is formed, is known in chemical terms as *transesterification*. It is noteworthy that in the course of two successive transesterification reactions between an enzyme and a DNA strand, in which the first reaction generates a break in the DNA strand and the second reaction reseals that break, the *Lk* value of the DNA ring may change while the DNA strand is transiently broken.

The postulated critical role of an enzyme hydroxyl group stemmed from the idea that there would be little loss of energy in the simultaneous breakage and formation of the same bonds between the protein and the broken DNA strand. Therefore, the resealing step could ensue without the help of another spontaneous reaction.[14] There was, however, no compelling reason to rule out other possibilities, such as the formation of a nitrogen-phosphorous (N—P) bond or a sulfur–phosphorous (S—P) bond. As it turned out, years would pass before the transesterification hypothesis was substantiated and the chemical nature of the reaction steps established.

CATCHING A STEALTH ENZYME IN ACTION

Relying on Sherlock Holmes's dictum requires that for a given event, all but one explanation be eliminated. Unlike the legendary detective, however, few scientists are confident enough, or sufficiently arrogant, to believe that they could come up with *all* possible explanations of a phenomenon. Therefore a critical test of the proposed transesterification mechanism became the identification of the predicted protein–DNA covalent intermediate. We summarize here some of the experiments that led to finding and characterizing the putative covalent intermediate in the reaction catalyzed by the *E. coli* enzyme; similar experiments with related enzymes are described in later chapters, especially Chapter 4, and some additional details are provided in Appendix 1.

The first break in that quest came about 5 years after the proposal of the transesterification mechanism. Richard Depew and Leroy Liu, then working in my laboratory, had joined the quest. We found that when a protein denaturant was added to a

14. For the relaxation of a supercoiled DNA, the overall reaction is energetically favorable because a supercoiled DNA is in a higher energy state than the same DNA in the relaxed form.

mixture of the *E. coli* ω protein and DNA (particularly a single-stranded DNA), a small amount of a protein–DNA complex having all the markings of a covalent intermediate would form.[15] A protein denaturant is a reagent that unfolds ("denatures") the intricate structure of a protein. In these experiments, sodium dodecyl sulfate (SDS), present in many household detergents, was one of the protein denaturants we used; alkali was another.[16] Alkali denatured the DNA as well as the protein, as the base pairs between DNA strands come apart when the bases T and G become negatively charged at a high pH.

In general, proteins that are noncovalently bound to DNA dissociate from it when they are denatured. We found, however, that a small fraction of the *E. coli* ω protein bound to single-stranded or negatively supercoiled DNA could withstand treatment with either detergent or alkali and remain tenaciously bound to the DNA. This finding suggested that, at a given moment during the normal reaction, the majority of DNA strands with bound ω protein would remain intact; however, a small fraction might be broken by the bound enzyme, with one end of each broken strand now covalently linked to the enzyme. In such a scenario, we reasoned, addition of a protein denaturant would rapidly denature the enzyme and prevent it from rejoining the broken DNA strand, thus "catching" the small fraction of DNA molecules with transient breaks.

Following these initial observations made in 1976–1978, our work over the next couple of years provided strong evidence that the trapped protein–DNA complex was indeed the covalent intermediate of the transesterification mechanism depicted in Figure 2-4. These experiments, conducted by Yuk-Ching Tse and Karla Kierkegaard, then graduate students in my laboratory, also revealed that the hydroxyl group of a tyrosyl residue of the enzyme apparently attacked a DNA phosphorus to form a 5′-phosphotyrosine bond, leaving a 3′-hydroxyl group at the broken end of the DNA strand (see Appendix 1 for details).

But could this trapped covalent complex be an artifact of adding the protein denaturant, rather than the true intermediate of the transesterification reaction? An answer came in 1986, when Yuk-Ching Tse-Dinh, by then at the E. I. du Pont Central Research and Development, took a clever approach in using a DNA oligonucleotide as short as eight nucleotides, instead of a long DNA strand, as a substrate in the cleavage reaction. Under these conditions, the broken DNA fragment (resulting from cleavage) that was not covalently attached to the enzyme could diffuse away, leaving behind the shortened oligonucleotide now covalently attached to the enzyme. In this reaction, formation of the covalent intermediate did not require denaturation of the enzyme, therefore the enzyme would retain its ability to rejoin this shortened DNA

15. Depew, R.E., et al. 1976. *Fed. Proc.* **35:** 1493; 1978. *J. Biol. Chem.* **253:** 511–518; Liu, L.F. and Wang, J.C. 1979. *J. Biol. Chem.* **254:** 11082–11088.

16. Some proteins, such as the keratins that form skin and hair, are highly resistant to protein denaturants; otherwise it would be disturbing if washing one's hair with a detergent like SDS would denature the hair as well as the scalp.

oligonucleotide covalently linked to it to another DNA strand having a 3′-hydroxyl terminus. This was in fact the case, demonstrating that the observed covalent complex is a true intermediate in DNA breakage and rejoining.[17]

Thus, the mechanism that I postulated in 1971 was on the right track.[10] In 1979 I coined the name "DNA topoisomerase" to denote enzymes that transiently break DNA backbone bonds to permit interconversion of topological isomers ("topoisomers"), such as the same DNA ring with different linking numbers; consequently, the *E. coli* ω protein became *E. coli* DNA topoisomerase I.[18,19] In 1991, Richard Lynn, a graduate student in my laboratory, identified the particular tyrosyl residue, Tyr-319, the 319 residues from the amino terminus of the enzyme, as the reactive tyrosyl group that attacks the DNA phosphorus to form a 5′-phosphotyrosine covalent link, leaving a 3′-hydroxyl group on the other end of the broken DNA strand. And it is this hydroxyl group which, in turn, attacks the phosphotyrosine bond in the subsequent step of rejoining the DNA strand.[20,21] By then, two decades had passed from the initial discovery of the *E. coli* ω protein to the final confirmation of its mechanism through the identification of the covalent enzyme–DNA link.

WHY IT TOOK SO LONG

In nature, the *Magicicada* takes 17 years to emerge from underground. But even the patience of a *Magicicada* would be tried during the long years it took to prove the proposed mechanism of action of the *E. coli* ω protein, today known as *E. coli* DNA topoisomerase I. Why did it take so long?

It is not uncommon in the world of academe for 5 or 6 years to pass between each significant advance in a long research quest; a graduate student often spends such a period on a Ph.D. thesis project. But the long duration of this particular quest also reflected the sea changes in the biological sciences that occurred during the two decades between the discovery of the enzyme and the final confirmation of its catalysis of DNA breakage and rejoining by transesterification. Reverse genetics, or the identification of a gene from its product, was a laborious practice in 1971, and the *topA* gene that encodes *E. coli* DNA topoisomerase I was not identified until 10 years

17. Tse-Dinh, Y.-C. 1986. *J. Biol. Chem.* **261:** 10931–10935.

18. Wang, J.C. and Liu, L.F. 1979. In *Molecular genetics,* Part III (ed. J.H. Taylor), pp. 65–88. Academic Press, New York.

19. Naming enzymes can be a bewildering task at times. For closely related enzymes of a particular organism, their numbering is typically done in the order of their discovery. Exceptions to this rule have often been made, however, and some of them will be described later.

20. The protein was encoded by 865 codons, although the first amino acid from its amino end was rapidly removed inside *E. coli*, leaving a polypeptide of 864 amino acid residues; rigorously speaking, Tyr-319 refers to counting from the first codon rather than from the amino terminus of the mature protein obtained from *E. coli* cells.

21. Lynn, R.M. and Wang, J.C. 1989. *Proteins* **6:** 231–239.

later.[22] The late 1970s marked the beginning of the era of cloning and recombinant DNA, and from then on the field of biomedical sciences would never be the same. For example, during the 1970s our studies of *E. coli* DNA topoisomerase I required hundreds of liters of *E. coli* culture. But by 1983, our cloning, with Kathy Becherer, of the *E. coli topA* gene that encoded DNA topoisomerase I enabled the preparation of large amounts of the enzyme from only a few liters of culture and without a lot of fuss.[23] The development of DNA sequencing techniques in the late 1970s also made it possible to determine (from DNA sequences) the amino acid sequences of the proteins encoded by genes. Before those advances it would have been impossible to determine that Tyr-319 of *E. coli* DNA topoisomerase I is the residue involved in DNA strand breakage and rejoining. Recombinant DNA methodology also made it an easy task to show, in 1991, that if Tyr-319 is replaced by some other amino acids (e.g., a phenylalanine or serine), then the activity of the enzyme is completely lost.[22]

MANY ROADS LEAD TO ROME

In the natural world, evolution often yields several distinctive solutions for a biological process. Because the problem of untangling DNA is deeply rooted in the plectonemic nature of the DNA double helix, it manifests itself not only in replication, but also in many other cellular transactions of DNA (which we shall discuss in Chapters 7–9). It is therefore no surprise that Nature has devised various alternative strategies to deal with the problems posed by DNA untanglement. *E. coli* DNA topoisomerase I represented the first strategy we uncovered, and a number of other DNA topoisomerases would soon be found in Nature's repertoire. The DNA topoisomerases now known are classified into four subfamilies (IA, IB, IIA, and IIB).[24] Type I enzymes break and rejoin one DNA strand at a time, whereas type II enzymes break and rejoin both strands of a DNA double helix in concert. Further division into A or B categories is based on their reaction mechanisms and amino acid sequences. All of these enzymes rely on the use of transesterification chemistry: They introduce transient breaks and always reseal these breaks before leaving the DNA. These enzymes are therefore stealth workers: They come and go, leaving no trace of having altered the chemical bonds in DNA. In their presence, DNA strands or double helices can move through one another as if the physical boundaries between them had disappeared. The beauty of the reactions catalyzed by DNA topoisomerases cannot be fully appreciated, however, without going through some of the details, which we shall consider in the next four chapters.

---

22. Trucksis, M. and Depew, R. E. 1981. *Proc. Natl. Acad. Sci.* **78:** 2164–2168; Sternglanz, R., et al. 1981. *Proc. Natl. Acad. Sci.* **78:** 2747–2751.

23. Wang, J.C. and Becherer, K. 1983. *Nucleic Acids Res.* 11: 1773–1790.

24. Wang, J.C. 1996. *Annu. Rev. Biochem.* **65:** 635–692.

CHAPTER 3

Breaking a DNA Strand and Holding on to the Broken Ends

"Alexander the Great would sneer. Twenty-three centuries after he slashed through the Gordian knot, mathematicians have finally made their first stab at figuring out how long it takes to untangle a tangle."

Charles Seife, reporting in News of the Week, *Science* **291:** 965, 2001

According to ancient Greek legend, whoever untied the intricate knot in the town square of Gordium would rule all of Asia. Then, in the year 333 B.C., came Alexander the Great. Noticing that the rope forming the knot had no discernible ends, the 23-year-old former student of Aristotle slashed it with a single stroke of his sword and declared the knot unraveled!

"Foul!" cried the mathematicians. Indeed, the great warrior had violated the cardinal rule for the time-honored knot-untying game: that two segments of a knotted ring are allowed to cross each other through a temporary break in one, but cutting the string to form two or more free ends is strictly forbidden. Twenty-three centuries later, however, any nonmathematician eager for an expert progress report on how long it might take to properly untie a knot would surely be disappointed, because the answer, in one word, is "eventually." By systematically manipulating strand crossings according to a set of rules while allowing the transient breakage of the string, the mathematicians have concluded that any knot can eventually be untied! Well, however unsatisfying such an answer might seem, the promise of eventual success is surely more encouraging than the possibility of struggling at a problem forever!

Yet decades before mathematicians set this new milestone in knot-untying, the *Escherichia coli* ω protein, later given the formal name *E. coli* DNA topoisomerase I, was found to be capable not only of relaxing a negatively supercoiled DNA, but also of tying and untying complicated knots in a single-stranded DNA ring![1] Here the DNA strand forming the knots has no ends of any kind, and the knots tied into it are therefore what mathematicians would certify as "true knots" rather than the "trivial knots" that can be tied in a linear strand. Unlike Alexander the Great, however, the enzyme apparently follows the same strict rules the mathematicians do. And unlike the gingerly moves of mathematicians, it can untie a knot in minutes, or perhaps even in seconds under favorable conditions! How does topoisomerase I accomplish such a remarkable feat? We shall return to this question after a discussion of a hallmark of this enzyme.

TYPE IA TOPOISOMERASES CAME ON STAGE

When the ω protein first made its debut in an *E. coli* cell extract in the late 1960s, it was thought to be a DNA endonuclease (see Chapter 2). Like Alexander's sword, the putative endonuclease was thought to cut one of the strands of a DNA double helix, but the cut would quickly be patched up by the robust action of DNA ligase in the same cell extract. In the case of a supercoiled DNA, it was then reasoned that the supercoils could escape during the time window between the breakage and resealing of a DNA strand.[2]

As we saw in Chapter 2, this interpretation was soon torpedoed by the finding that DNA ligase played no part in the observed supercoil removal. By 1971 it was clear that the *E. coli* ω protein, or DNA topoisomerase I (Top1, for short), relaxes a supercoiled DNA by accomplishing both the breakage and rejoining in a most unexpected way.[3] The Top1 of *E. coli* turned out to be the founding member of a ubiquitous family of enzymes now termed the type IA DNA topoisomerases.

In many ways, among all DNA topoisomerases, the workings of the type IA subfamily are probably the most difficult to decipher and comprehend. The strong preference of these enzymes for negatively but not positively supercoiled DNA rings[3] and their ability to tie or untie knots in a single-stranded DNA ring[1] are some of the peculiar properties that set them apart from the other subfamilies.

There is substantial evidence that the type IA enzymes act by a strategy termed the "enzyme-bridging" mechanism: When a type IA DNA topoisomerase transiently cleaves a DNA strand, it holds on to both of the broken ends it has created. In other words, the enzyme molecule is itself in direct physical contact with both broken ends, and it forms a bridge across the break it has created in the DNA strand until it subsequently rejoins the strand. How does it manage this ambidextrous act?

Even more difficult to comprehend is that the enzyme not only promotes the breakage and rejoining of the DNA strand, but that while holding both broken ends

1. Liu, L.F., et al. 1976. *J. Mol. Biol.* **106:** 439–452.
2. Wang, J.C. 1969. *J. Mol. Biol.* **43:** 263–272.
3. Wang, J.C. 1971. *J. Mol. Biol.* **55:** 523–533.

of the strand, the enzyme must also allow, or perhaps facilitate, the movement of another DNA strand through the gap between the two broken ends, otherwise no net topological change would occur in a DNA ring.

In this chapter we shall consider some of the key mechanistic issues relating to the type IA DNA topoisomerases: the lines of evidence leading to their classification as type I topoisomerases, which are by definition enzymes that break DNA strands one at a time; the mechanism by which a type IA enzyme can tie different "knots" in single-stranded DNA rings or remove such knots; and the experimental evidence supporting the enzyme-bridging model. However, we will not discuss the actual functions of these enzymes inside living cells until Chapters 7 and 8. It suffices to say here that these enzymes play vital roles in nearly all living organisms, and without them these organisms would have most pathetic lives before their eventual demise.

A CURIOUS SPECIFICITY FOR NEGATIVELY SUPERCOILED DNA

The idea that *E coli* DNA topoisomerase I breaks one DNA strand at a time can be traced to the first identification of this enzyme. The 1971 report on the *E. coli* ω protein included an observation that nearly undermined the entire paper after its submission to the *Journal of Molecular Biology* in July of 1970: the enzyme specifically removed negative but not positive supercoils.[3]

The experiments investigating the action of the ω protein on a positively supercoiled DNA were originally conceived with one specific purpose in mind: A demonstration that the protein could remove both positive and negative supercoils would strengthen the argument that it transiently broke DNA strands. Why? Because any alternative interpretation invoking an unwinding of the DNA double helix to offset the negative supercoils, such as a change of the DNA helical geometry from the B- to the A-helix structure or the disruption of base pairing in a short stretch of the DNA, would not account for the removal of positive supercoils. Yet the experiments proved just the opposite: Positively supercoiled DNA was completely refractory to the enzyme!

This unanticipated finding called into question the proposal that the enzyme acted by transiently breaking DNA strands. The two anonymous reviewers of the paper in which this proposal was made,[4] probably already bewildered by the seemingly bizarre activity never before observed, found the proclaimed specificity of the protein disturbing. If the enzyme catalyzed the removal of negative supercoils by transiently cleaving a DNA strand, they questioned, why did it not remove positive supercoils in the same way? How could an enzyme distinguish supercoils of different sign?

---

4. One of the reviewers turned out to be Bruce Alberts, a noted biochemist who later served two terms as president of the U.S. National Academy of Sciences. Alberts told me many years later that his first impulse after reading the manuscript was that the results were too bizarre to be credible, and that the manuscript should be rejected outright; he changed his mind after reading the manuscript again 2 weeks later.

The curious specificity of the ω protein for negative supercoils led me to postulate that the enzyme acted on one strand of a DNA double helix; that is, it must first disrupt a short stretch of the DNA double helix before it could cut a strand.[3] Because a negatively supercoiled DNA ring is underwound, it seemed likely that its DNA double helix would be more receptive to the uncoiling of a short DNA segment than would the helix in a relaxed DNA. By the same logic, the DNA double helix in a positively supercoiled DNA ring, already in an overwound state, would resist more fiercely the formation of an unpaired single-stranded bubble. Later studies would provide strong evidence that opening a single-stranded bubble is indeed much easier in a negatively supercoiled than in a positively coiled or relaxed DNA ring, but in 1970 the idea that negative supercoiling facilitated the disruption of base pairing was not widespread.[5]

A particular technical issue also worried the reviewers of the 1970 manuscript. At that time, negatively supercoiled DNA rings could either be isolated from natural sources, or made in the laboratory by a method introduced a year earlier. In this approach (shown in Fig. 3-1), negatively supercoiled DNA is digested very lightly with an endonuclease called pancreatic DNase I, which more or less randomly nicks one of the DNA strands in a double-stranded DNA to produce nicked DNA rings. To reduce the total number of times the two strands revolve around each other in the nicked DNA ring, a chemical called ethidium is added to the DNA[6] (see below). Following treatment with DNA ligase to seal all of the nicks introduced by DNase I, the bound ethidium molecules are removed, such as by extracting the aqueous solution with butanol (butyl alcohol) to yield a DNA ring negatively supercoiled to a predetermined extent. This sequence results in a family of DNA rings negatively supercoiled to different degrees.[6] The DNA rings produced with this method can be as much as two to three times more negatively supercoiled than those isolated from cells—which as we shall see turns out to be a real advantage.

5. The first demonstration that negative supercoiling favors DNA strand separation was reported in 1968 (Vinograd, J., et al. 1968. *J. Mol. Biol.* **33:** 173–197), and an account of the energetics of winding and unwinding of the DNA double helix in a supercoiled DNA was reported in 1970 (Bauer, W. and Vinograd, J. 1970. *J. Mol. Biol.* **47:** 419–435).

6. The compound ethidium, often as its bromide salt, was used in the treatment of sleeping sickness in cattle, an African trypanosome infection transmitted by the tse-tse fly. The compound was introduced to Jerome Vinograd's laboratory by Jean-Bernard LePecq in 1965, the year supercoiled DNA was discovered there. While a graduate student with Claude Paoletti in France, LePecq had carried out a detailed study of ethidium binding to DNA. He intended to do postdoctoral studies at Stanford University, but in 1965, a shortage of laboratory space there landed him at Caltech, in Norman Davidson's laboratory, adjacent to Vinograd's. Ethidium bromide became an instant hit there, and soon gained worldwide popularity among those working with DNA. In addition to its use in the study of DNA supercoiling, binding of this chemical to DNA produces a bright orange-red fluorescence under ultraviolet light, making it a convenient reagent for detecting and quantifying small amounts of DNA. The mode of ethidium binding to DNA by insertion between base pairs was termed "intercalation" by Leonard S. Lerman (*J. Mol. Biol.* 1961. **3:** 18–30). For many years the intercalation of ethidium ino a DNA was believed to unwind the double helix by 12°. In 1974, however, improvements in experimental measurements and data analysis finally showed that the unwinding angle was actually 26° (Wang, J.C. 1974. *J. Mol. Biol.* **89:** 783–801).

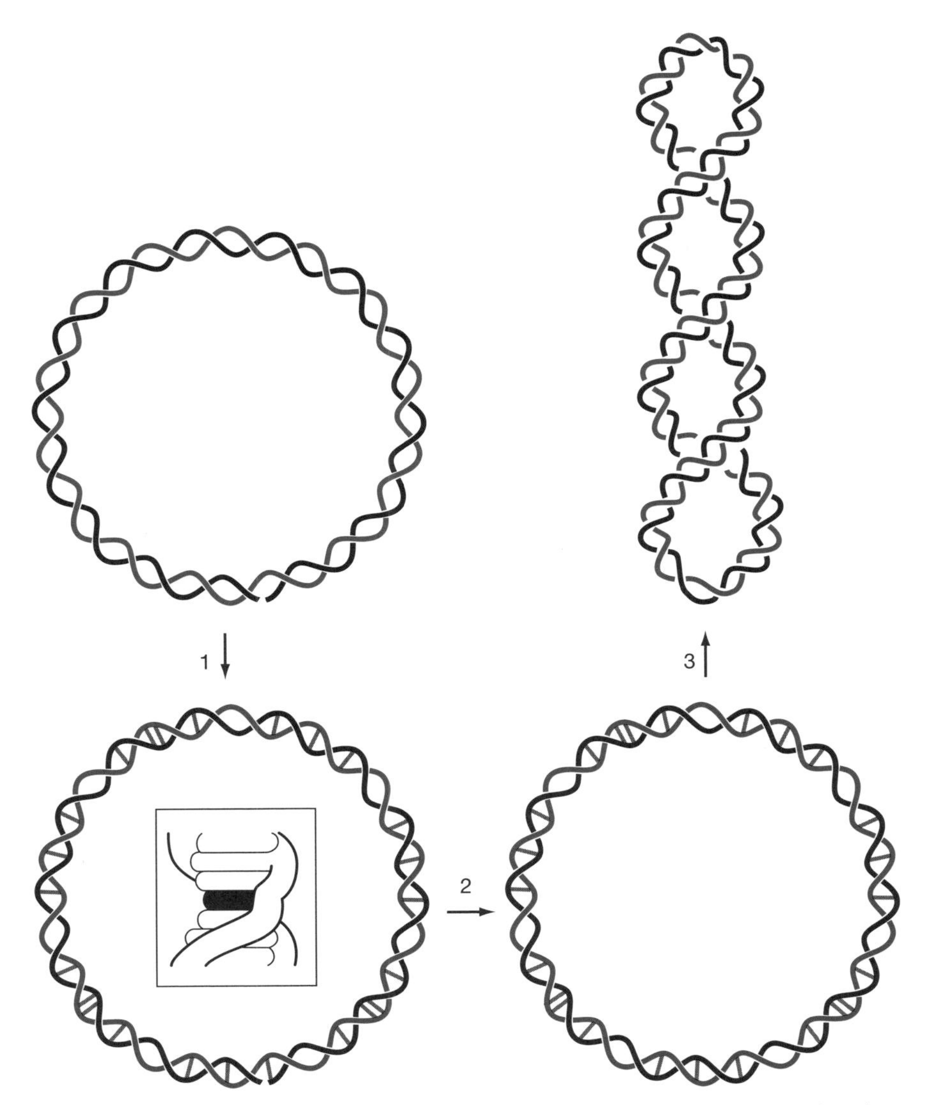

Figure 3-1. Preparation of DNA rings negatively supercoiled to different extents. A nick is depicted at the bottom of the DNA ring in the *upper left*. In Step 1, ethidium, added to the DNA, is depicted by gray bars perpendicular to the helical axis of the DNA ring. The *inset* represents an enlarged view of one intercalated ethidium (shown as a black disk) in a short DNA segment. Because of the presence of the nick, unwinding of the DNA by ethidium does not lead to contortion of the entire DNA ring. In step 2, DNA ligase is used to seal all nicks in the DNA, fixing the *Lk* of the DNA ring. In Step 3, DNA-bound ethidium molecules are removed. The DNA becomes underwound (negatively supercoiled) and coils up in space (*upper right*). Drawings of the DNA rings are taken, with permission, from Wang, J.C. 1982. *Sci. Am.* **247:** 94–109.

Ethidium, known as a DNA "intercalator," is a planar molecule that binds to a DNA double helix by inserting itself between two adjacent planar base pairs (see inset in the lower left DNA ring in Fig. 3-1). This insertion, more specifically termed "intercalation," uncoils the DNA by an angle of 26°.[6] Thus, 14 (360°/26°) bound ethidium molecules would uncoil the DNA by a full turn. If the nicks in a DNA ring are sealed by DNA ligase in the presence of a certain amount of bound ethidium molecules, then the linking number of the resulting DNA is reduced relative to that of the same DNA ring sealed in the absence of ethidium by a value equal to one-fourteenth of the total number of ethidium molecules bound to each DNA ring.

No reagent was then known to wind the DNA double helix more tightly rather than unwind it, however, and positively supercoiled DNA rings could therefore not be made in a similar way.[7] For this reason a relaxed DNA ring in the presence of ethidium was used to test the activity of the *E. coli* ω protein on positively supercoiled DNA. Uncoiling the double helix in a relaxed DNA ring by ethidium binding would be expected to induce compensatory positive supercoiling of the DNA ring.[8] But the reviewers remained unconvinced, asking whether the very presence of ethidium, rather than the positive supercoiling of the DNA ring per se, prevented the enzyme's relaxation of a positively supercoiled DNA.

To rule out such a trivial explanation, an additional experiment was performed. Using the trick of sealing a nicked DNA in the presence of ethidium, I prepared a DNA sample that was negatively supercoiled to a very high degree, as well as another sample of completely relaxed DNA, made by sealing the same nicked DNA in the absence of ethidium. I then purified the samples to remove all bound ethidium and added back a calculated amount of ethidium to each of the two DNA samples. After

7. It would not be until 1983 that the binding of a compound known as netropsin was found to tighten the DNA helix (see Snounou, G. and Malcom, A.D. 1983. *Mol. Biol.* **167:** 211–216). In the same year, positively supercoiled plasmid DNA was isolated from *E. coli* cells treated with novobiocin (Lockshon, D. and Morris, D.R. 1983. *Nucleic Acids Res.* **11:** 2999–3017; how this treatment leads to positive supercoiling of an intracellular plasmid is discussed in Chapter 8). Later, it would also become possible to prepare positively supercoiled DNA samples by use of the enzyme reverse gyrase (Kikuchi, A. and Asai, K. 1984. *Nature* **309:** 677–681), or with a combination of RNA polymerase and *E. coli* DNA topoisomerase I (Liu, L.F. and Wang, J.C. 1987. *Proc. Natl. Acad. Sci.* **84:** 7024–7027).

8. Those unfamiliar with DNA supercoiling are often confused by the sign of supercoiling of a covalently closed DNA ring when its helical geometry is altered by ethidium binding. It is helpful to keep in mind that as increasing numbers of ethidium molecules bind to a covalently closed DNA ring, the linking number, Lk, remains unaltered (because Lk can change only by transient breakage of at least one DNA strand) while Lk^0 continuously decreases (see Fig. 3-1). Thus, if DNA is initially negatively supercoiled and by definition $\Delta Lk \equiv (Lk - Lk^0) < 0$, then as Lk^0 continuously decreases with the binding of increasing numbers of ethidium molecules to the DNA, the value of ΔLk first becomes less and less negative, and then becomes increasingly positive after passing through the point $\Delta Lk = 0$, at which the DNA is completely relaxed.

this addition, the highly negatively supercoiled DNA ring remained negatively supercoiled, albeit to a lesser extent than initially as the result of ethidium binding, but unwinding of the DNA double helix by binding of the same amount of ethidium to the relaxed DNA ring made the ring positively supercoiled. The results of this experiment showed that ethidium itself did not inhibit the relaxation of the negatively supercoiled DNA, and that its linking number was increased by treating the DNA with the ω protein. However, the linking number of the positively supercoiled DNA in the presence of the same amount of ethidium remained unaltered upon incubation with the ω protein, thus confirming the earlier conclusion that the protein could not remove positive supercoils. Following a few more exchanges through the journal editor, the reviewers finally relented, and the article was accepted on November 2, 1970 and published in early 1971.[2]

Years would pass, however, before a more direct demonstration would show that the specificity of the enzyme is indeed tied to its opening of a single-stranded bubble in a DNA. By making a DNA ring containing a small single-stranded region (in this case the nucleotide sequences of the strands within a short stretch of the DNA ring were deliberately made noncomplementary to prevent base pairing between them), we showed that as long as a DNA ring contains a short single-stranded region in one of its strands, it can be readily relaxed by *E. coli* DNA topoisomerase I even when the DNA is made into a positively supercoiled form by ethidium binding.[9]

BREAKING ONE DNA STRAND AT A TIME

In Fig. 2-4 of the preceding chapter, an *E. coli* enzyme is depicted to break only one of the two strands of a double-stranded DNA. The specificity of the enzyme for negatively supercoiled DNA and the mechanism proposed to account for this specificity provided the initial evidence that the enzyme transiently breaks one DNA strand at a time. More direct evidence for this came soon after the identification of the enzyme–DNA covalent intermediate (described in Chapter 2 and Appendix 1). Starting with a negatively supercoiled DNA ring, we showed that formation of the covalent intermediate yields DNA rings with only one of their strands nicked; even better, we found that formation of the covalent intermediate between an *E. coli* Top1 molecule bound to a single-stranded DNA ring converts the ring to a linear molecule of the same length but with a protein linked to one of its two ends. The idea that the enzyme breaks one strand at a time is also consistent with its ability to tie or untie knots in a single-stranded DNA ring, which we had alluded to at the beginning of this chapter.

9. Kirkegaard, K.A. and Wang, J.C. 1985. *J. Mol. Biol.* **185:** 625–637.

TYING KNOTS IN A SINGLE-STRANDED DNA RING

We now return to the question on the knotting and unknotting of a single-stranded DNA ring by *E. coli* Top1. Historically, the first DNA found to be in the form of a ring was not the double-stranded polyoma DNA that was discussed in Chapter 1, but the DNA of a small *E. coli* virus called ϕX174.[10] Viral ϕX174 DNA is a peculiar member of the DNA world: It contains only one DNA strand rather than a pair of intertwined strands of complementary sequences. Upon its entrance into an *E. coli* cell, however, the single-stranded ϕX174 DNA ring is soon converted to a double-stranded form through the synthesis of a strand complementary to the original viral DNA. Many copies of the original single-stranded viral DNA are then made from this double-stranded DNA template, and these copies are packaged into viral particles for infecting more *E. coli* cells.

The discovery of ϕX174 was soon followed by the identification of other phages containing single-stranded DNA rings, including a phage called fd. In 1976, Leroy Liu and Richard Depew, then working in my laboratory, found that treatment of single-stranded fd DNA with *E. coli* Top1 yielded a novel product.[1] When spun in an alkaline medium in an ultracentrifuge, this product sedimented as a heterogeneous species with velocities from 20%–50% greater than that of the original single-stranded DNA ring. The pH of the alkaline medium used in this sedimentation analysis was greater than 12. At such a high pH, pairing between the bases in the single-stranded fd DNA is completely disrupted—many of the DNA bases become negatively charged, and charge repulsion would push apart any base-paired strands. It was therefore highly unlikely that the 20%–50% increase in the average sedimentation velocity of the treated fd DNA ring could be attributed to folding of the single-stranded DNA through base pairing.

The increase in the sedimentation velocity also appeared to have little to do with residual protein molecules that might remain associated with the DNA. The sedimentation pattern of the DNA was unaffected by all treatments that could removed DNA-bound proteins, such as extracting the *E. coli* Top1-treated fd DNA with the solvent phenol, which dissolves most proteins, or digesting the treated DNA exhaustively with a proteinase, an enzyme that breaks down proteins into short peptides.

Most remarkably, we found that a single cleavage of the Top1-treated fd DNA by an endonuclease would convert it to a form with a uniform sedimentation velocity in the same alkaline medium. Furthermore, this linear DNA product was in all respects indistinguishable from a linear fd DNA formed by a single cleavage of an untreated fd DNA ring. How could such results be explained? A DNA sample exhibiting a broad distribution in sedimentation velocity is usually indicative of a heterogeneous collection of species with varying sedimentation velocities. Because this apparent heterogeneity completely disappeared with a single cleavage of the DNA ring, we reasoned that the crucial issue here could well be the topology of the ring: Perhaps treatment

10. Fiers, W. and Sinsheimer, R.L. 1962. *J. Mol. Biol.* **5:** 424–434.

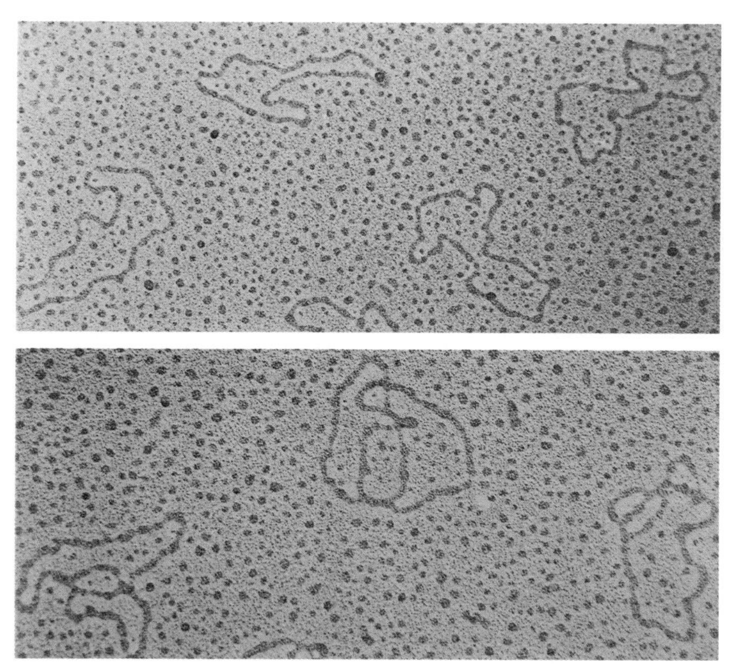

Figure 3-2. Two electron micrographs of phage fd DNA. The *upper* micrograph shows untreated fd molecules and the *lower* one shows fd molecules after treatment with the *E. coli* ω protein followed by extensive treatment to remove any protein. Reprinted, with permission, from Liu, L.F., et al. 1976. *J. Mol. Biol.* **106:** 439–452.

of the single-stranded DNA ring with the enzyme had introduced knots of varying complexities into it. This would explain why a single cleavage of any DNA ring in the heterogeneous collection would convert it to the same linear product. Examining the treated sample under an electron microscope supported our interpretation: DNA rings with a knotted appearance were indeed seen after, but not before, treatment of fd DNA with *E. coli* Top1 (Fig. 3-2).[1]

Figure 3-3 illustrates how knots of different complexities can be tied in a single-stranded DNA ring by an enzyme that breaks and rejoins a DNA strand. The linking of two single-stranded loops separated by intramolecular base pairing between two distal regions in a single-stranded DNA ring, can occur by the transient breakage of one of the loops and passage of the other loop through the transient break. As illustrated in Figure 3-3, if the loops are separated by a single intertwining with one another, then their linkage could form a trefoil, the simplest knot. A more complex twist knot would form, however, if the duplex divide contains multiple intertwines. Even in a single-stranded DNA, nearly half of the nucleotides are typically base-paired in the Watson–Crick fashion at or around room temperature in a buffer containing an

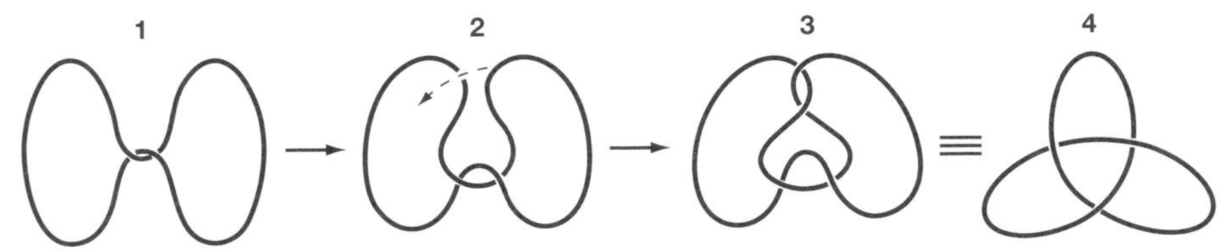

Figure 3-3. A mechanism of knotting single-stranded DNA rings by the *E. coli* ω protein (Top1).[1] (*1*) Two segments of a single-stranded DNA ring twist around each other once, dividing the ring into two loops. Base pairing between short stretches of nucleotides of complementary sequences along the single-stranded DNA ring can form right-handed intertwines. (*2*) Top1 makes a transient break in the loop on the left, allowing passage of the right loop through it; locking of the two loops after resealing of the break gives the structure shown in *3*. (*3*) The structure shown is topologically identical to that shown in *4*, which is called a trefoil. The trefoil depicted here, and its mirror image, are the simplest knots. More complex knots would form with a longer twisted region or regions, or when there are more loops and more than one strand-passage event between them.

appropriate concentration of counterions (base pairing in other ways also occurs, but to much lesser extents). These base-paired regions usually consist of short stretches of adventitious base pairs with or without interspersed base mismatches, which can be in the form of a single-stranded loop or a "bubble" with two unmatched strands.

The knotting scheme we have just considered explains how treatment of a single-stranded DNA ring with an enzyme that breaks and rejoins a DNA strand could yield a very heterogeneous knotted product. Owing to the presence of many differently folded structures in a single-stranded DNA ring, different knots are produced, and subsequent disruption of base pairing in a high–pH medium cannot alter the types of knots. Such a mechanism also suggested a further test for this knotting scheme. Because of their short lengths and imperfections, duplex regions in a DNA single strand that are held together by the adventitious pairing of short sequences are much less stable relative to paired complementary strands in a long DNA double helix. Their stability is also sensitive to the concentration of positively charged counterions in a solution containing the DNA: In a neutral aqueous solution, the phosphate groups in the DNA backbones are negatively charged, and the charged backbones would cause any short base-paired region in a DNA strand, or even in two complementary strands of a long DNA double helix, to come apart if there were no counterions to shield the negatively charged phosphate groups. Therefore, treatment with *E. coli* Top1 in media containing decreasing concentrations of counterions should produce knots of reduced complexities. This was confirmed when we found that in the treatment of fd DNA with *E. coli* Top1, the mean sedimentation velocity of the knotted product decreased with decreasing counterion concentration. Furthermore, we found that if a knotted product that had first been formed by the *E. coli* enzyme at a high counterion concentration was again treated with the same enzyme, this second treatment decreased the mean sedimentation velocity of the knotted product in the presence of a decreasing counterion concentration in the medium. The *E. coli* enzyme could therefore not only tie complex knots in a single-stranded DNA ring, but it could also untie them under a different set of reaction conditions.[1]

SUBTLER THAN ALEXANDER THE GREAT YET FASTER THAN THE MATHEMATICIANS

Knotting and unknotting problems have attracted the interests of many mathematicians for centuries, precisely because such seemingly simple problems have proven to be enormously challenging. The style of *E. coli* DNA topoisomerase I in its knotting and unknotting of DNA rings is much subtler than the rash manner of Alexander the Great. The topoisomerase not only leaves the DNA strand intact after it has tied or untied a knot, it also leaves unaltered the order of arrangement of all nucleotides (the nucleotide sequence) along the contour of each fd DNA ring. We showed that the genetic information in an fd DNA ring remained unaltered after the knotting reaction by *E. coli* Top1: Once a knotted fd ring gained entry into an *E. coli* cell, it was as successful as its unknotted counterpart in producing healthy fd viruses.[1] Had there been any reshuffling of the nucleotide sequence, the genetic information would be compromised, and the production of functional viral particles would be impossible.

When we think a bit more about it, such a feat is absolutely amazing: An enzyme molecule, like a very nearsighted person, can sense only a small region of the much larger DNA to which it is bound, surely not an entire DNA ring. How can the enzyme manage to make the correct moves, such as to untie a knot rather than make the knot even more tangled? How could a nearsighted enzyme sense whether a particular move is desirable or undesirable for the final outcome?

In such a case the chore of the enzyme is helped by two phenomena in nature. First, there are random molecular movements at ordinary temperatures, termed Brownian motion (named after the naturalist Robert Brown, 1773–1858). It is this Brownian or thermal motion that keeps all of the molecules in a system in a state of constant agitation at ordinary temperatures. Second, Nature also provides a strong incentive for a molecule, or a collection of molecules, to move from a higher to a lower energy state.[11] It is most likely that this inherent bias in nature sets a preferred directionality toward the knotting or unknotting of an agitated DNA ring, depending on the reaction conditions. By permitting the passage of DNA strands through each other, a topoisomerase simply makes it possible for the topology of a DNA ring to change. A general principle in chemistry is that a catalyst promoting a reaction in one direction must also promote the same reaction in the reverse direction. An enzyme capable of tying a knot in a DNA ring must therefore also be able to untie it. It is the energetics of folding a single-stranded DNA ring under a given set of reaction conditions that determines what kinds of knots, or whether an "unknot"—a simple

11. This point was alluded to in the discussion of spontaneous and nonspontaneous reactions in Chapter 2.

ring without a knot in it—is formed in the presence of *E. coli* Top1 or some other type IA DNA topoisomerase.

TWO WAYS TO TIE OR UNTIE A KNOT

Our discussions have thus far left two questions untouched. First, in an ionic medium resembling that of the cellular milieu, the presence of *E. coli* Top1 has been found to convert an unknotted single-stranded DNA ring to a knotted one. Why, then, are the single-stranded ϕX174 or fd DNA rings extracted from the viral particles found exclusively in the unknotted form? Second, if an *E. coli* DNA topoisomerase I can tie or untie a knot in a single-stranded DNA ring by first breaking the strand, how are the two broken ends prevented from flying apart, owing to Brownian motion, and never finding each other again?

The answer to the first question likely lies in the state of a single-stranded DNA ring inside a cell. There, a single-stranded DNA is rarely "naked"; a protein that specifically binds to the DNA usually decorates it along its contour. And when a single-stranded DNA is packaged into a viral particle, it is associated with multiple viral protein molecules. In other words, adventitious base pairing in a DNA single strand is most likely minimal inside a cell, and therefore no knotting would occur, whether or not the DNA is accessible to a type IA DNA topoisomerase.

As to the second question, the answer a priori could lie either in the structure of the DNA molecule in this reaction or in an intrinsic characteristic of the enzyme. One way of cleaving a DNA such as that of fd without having the broken ends fly apart is to break the DNA only within a double-stranded region. In this case, the intact strand would hold the two broken ends in close proximity until they are subsequent rejoined. An alternative is for the enzyme itself, rather than the DNA, to hold the two broken ends together. The latter case is termed the "enzyme-bridging model" and is illustrated in Figure 3-4.[12] We saw in Chapter 2 that the 5′-phosphoryl end of the strand cleaved by *E. coli* Top1 is covalently anchored to Tyr-319 of the enzyme. The enzyme-bridging model requires that the other broken DNA end, containing the 3′-hydroxyl group, also be tightly bound to the enzyme; the enzyme would thus bridge the two ends of the DNA strand that it cleaves.

A structural prediction of this model is that each of the two enzyme regions that hold the broken DNA end must enclose a "hole" large enough to allow a passing DNA strand to sail under the protein bridge. The enzyme shown in Figure 3-4 is represented by a C-clamp; when it breaks a single DNA strand, one tip of the C becomes covalently attached to the 5′-phosphoryl group of the broken DNA end, whereas the other remains tightly associated with the 3′ side of the broken strand. We see that the C-clamp tips, by using a flexible hinge region in the enzyme, could widen the opening

12. The idea of a topoisomerase holding the DNA ends that it has generated originated with studies of the type II DNA topoisomerases (Brown, P.O. and Cozzarelli, N.R. 1979. *Science* **206:** 1081–1083; Liu, L.F., et al. 1980. *Cell* **19:** 697–707; see Chapter 4).

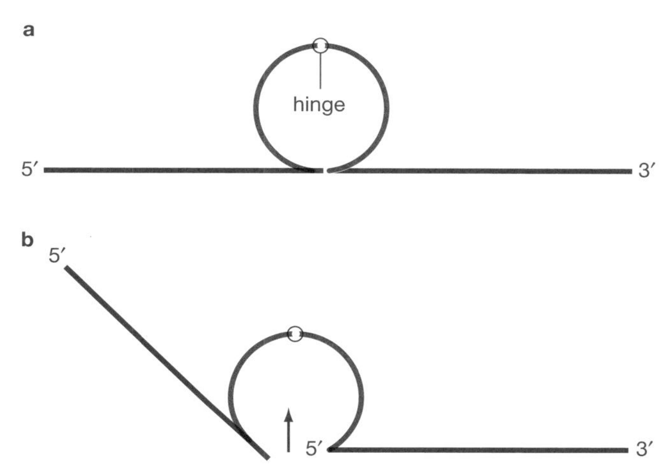

Figure 3-4. The enzyme-bridging model for a type IA DNA topoisomerase. (*a*) The enzyme is represented by a C-clamp with a hinge; for clarity, ony the scissile DNA strand is shown, and the enzyme is depicted to interact with only this DNA strand. (*b*) In the enzyme-bridging model, the enzyme remains associated with both ends of the broken DNA strand after it cleaves the strand and widens the strand interruption.

in the broken strand to allow the passage of another strand into the enzyme (we shall discuss further mechanistic details later in this chapter).

Evidence in support of the enzyme-bridging model came in 1985, when Karla Kierkegaard, then a graduate student in my laboratory, prepared special DNA substrates to direct the binding of an *E. coli* DNA topoisomerase I molecule to a particular region of a DNA. We found that the presence of a single-stranded loop in the middle of a double-stranded DNA directs the binding of an *E. coli* DNA topoisomerase I molecule to the region containing this loop. A method termed "DNA footprinting"[9] (see below) allowed us to use this substrate to obtain more detailed information pertaining to the contacts between the enzyme and DNA. The footprints of *E. coli* Top1 bound to various DNA fragments containing interior single-stranded loops showed that a typical bound enzyme interacts with both the 5′ and 3′ ends of the broken DNA.[9]

LINKING AND UNLINKING OF DNA RINGS

In 1978, Karla Kierkegaard and I demonstrated that *E. coli* DNA topoisomerase I could catalyze the formation of a double-stranded DNA ring from a pair of single-stranded component rings.[13] This outcome was actually entirely predictable and so did not pro-

13. Kirkegaard, K.A. and Wang, J.C. 1978. *Nucl. Acids Res.* **5:** 3811–3820; the same reaction had also been shown a year earlier for a type IB DNA topoisomerase (Champoux, J.J. 1977. *Proc. Natl. Acad. Sci.* **74:** 5328–5332).

In a footprinting experiment, a protein–DNA complex and the same DNA without the protein are each lightly treated with a reagent, usually a DNase or a chemical that randomly breaks DNA backbone bonds. Random cleavage of the DNA is usually carried out to the extent that approximately one break would be introduced in any stretch of DNA a couple of hundred base pairs in length. Through the use of DNA sequencing methods, the exact positions of all such cleavage sites within a particular region of a population of DNA molecules can then be precisely mapped. In the absence of the protein, all DNA cleavage sites are expected to be more or less evenly distributed along the DNA (depending on how "randomly" the reagent acts). In the presence of bound protein, however, the pattern of cleavage of the DNA by the same reagent is expected to be significantly altered. In the simplest case, when the binding of a protein to a particular region completely shields the region it occupies from cleavage by the reagent, the bound protein would leave a clean "footprint" in the DNA cleavage pattern: from heel to toe within this footprint, no DNA cleavage would occur. More often, however, the footprint is revealed by changes in the cleavage pattern in the absence of a bound protein, rather than by a complete absence of cleavages within the protein-bound region.

vide much new mechanistic insight: Two unlinked single-stranded DNA rings of complementary sequence represent a special case of a highly negatively supercoiled DNA ring—with an Lk of 0 and thus a difference between Lk and Lk^0 that is equal to $-Lk^0$. This reaction is therefore simply a special case of the relaxation of a highly negatively supercoiled DNA. Nevertheless, it nicely demonstrates that in the presence of a type IA DNA topoisomerase, DNA strands of complementary sequence can pair in the absence of any molecular ends. As we shall see in Chapter 7, this concept is relevant to recombination involving intracellular DNA.

More interesting mechanistically is the reaction involving catenation and decatenation of double-stranded DNA rings by the *E. coli* enzyme. Linking and unlinking of double-stranded DNA rings was first observed in reactions catalyzed by the type II DNA topoisomerases (to be described in Chapter 5). Shortly afterward, Yuk-Ching Tse in my laboratory and Pat Brown, working with Nick Cozzarelli at the University of Chicago, observed that a type IA DNA topoisomerase could also link or unlink double-stranded DNA rings.[14] The type IA reaction differs from that catalyzed by a type II DNA topoisomerase, however, in that in the former case at least one of the reacting pair of DNA rings must contain at least one preexisting break (a "nick") in one of its strands. Other than this difference, the linking of double-stranded rings by the type I and type II enzymes is similar in that neither enzyme requires the presence of homologous sequences in the participating DNA rings.

14. Tse, Y.C. and Wang, J.C. 1980. *Cell* **22:** 269–276; Brown, P. and Cozzarelli, N.R. 1981. *Proc. Natl. Acad. Sci.* **78:** 843–847.

The simplest interpretation of the catenation/decatenation reaction by a type IA enzyme is that the enzyme transiently breaks one strand of the nicked DNA ring at a position directly opposite the preexisting nick. The enzyme may favor that particular location because of its preference for an unpaired region in a double-stranded DNA: The presence of a nick would perhaps make it easier for the enzyme to disrupt base pairing at that site. By breaking the strand opposite the nick and holding onto the broken ends, the enzyme would in fact create a double-stranded interruption in the DNA through which another DNA double helix could pass.[14] We shall consider the mechanistic significance of this reaction in more detail in a later section of this chapter.

A C-CLAMP CAME INTO VIEW

One prediction of the enzyme-bridging model is the presence of a recess within *E. coli* Top1 that is large enough to hold at least one DNA strand or perhaps a DNA double helix that has just passed through the enzyme-mediated opening in the scissile strand (the DNA strand cleaved or to be cleaved by the enzyme). Visualization of this prediction would depend on three-dimensional structural studies of the enzyme. In the early 1990s, my collaborators Christopher Lima and Alfonso Mondragón, obtained crystals of a large fragment of *E. coli* DNA topoisomerase I, spanning amino acid residues 2–596 of the enzyme. X-ray diffraction studies of the crystals were soon completed and revealed a striking four-domain structure enclosing a large opening in its center.[15] The three-dimensional structure of the polypeptide chain, shown in Figure 3-5, is represented by a "ribbon model." Here the two prominent structural elements found in most proteins, the α-helices and β-strands, are represented, respectively, by short helical segments and short stretches of flat ribbons. In an α-helix, the polypeptide forms a right-handed helix that typically has 3.6 amino acid residues per turn and a helical pitch of 0.54 nm. In a β-strand, the polypeptide is in a more extended conformation, and adjacent β-strands often form a pleated sheet in which the β-strands can run either in the same direction (a parallel β-pleated sheet) or in opposite directions (an antiparallel β-pleated sheet).[16] The structure in Figure 3-5a comprises 18 discernible α-helices and a number of β-pleated sheets formed by more than a dozen β-strands (the polypeptide chain of the *E. coli* Top1 fragment depicted here is disrupted at several places because the amino acid residues in those regions are not sufficiently ordered in the particular crystals to allow their detection by X-ray diffraction). Several additional crystal structures of type IA DNA topoisomerases have since been determined, including crystals of a full-length *Thermotoga maritima* Top1 and *E. coli* DNA topoisomerase III (Top3); both of

15. Lima, C.D., et al. 1994. *Nature* **367:** 138–146.

16. The α-helix and β-pleated sheet structures of polypeptides were proposed by Linus Pauling and Robert Corey in 1951, on the basis of their extensive chemical knowledge and through model building. These and the double-helix structure of DNA are among the most prominent structural motifs in biological macromolecules.

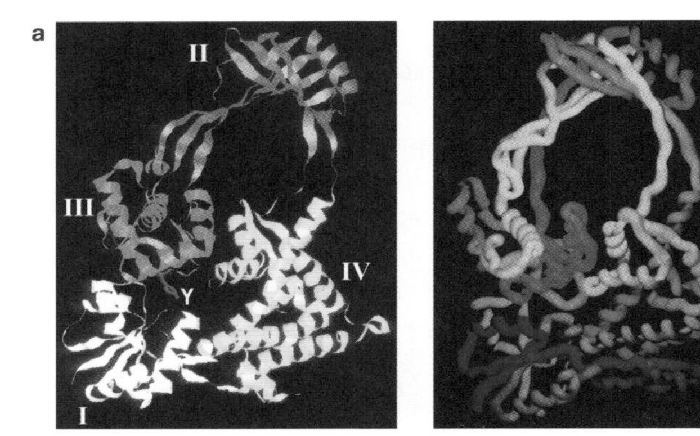

Figure 3-5. Folding of the large polypeptide fragment of *E. coli* DNA topoisomerase I (amino acid residues 2–596). (*a*) A ribbon model is shown. The four distinct domains are depicted in white (I), blue (II), green (III), and yellow (IV); Y marks the position of the active-site tyrosyl residue Tyr-319 (in red). (*b*) The same polypeptide is represented by a coil colored from blue to red along its contour. See text for details.

these type IA DNA topoisomerases closely resemble *E. coli* Top1[17,18] and an *Archaeoglobus fulgidas* type IA enzyme called "reverse gyrase" (to be described in a section at the end of this chapter).[19] All of these structures show the basic four-domain structure enclosing a large central hole with a diameter of about 2.5 nm, although the reverse gyrase of *A. fulgidas* shows a significantly smaller hole with a diameter of 1.6 nm.

The four-domain structure depicted in Figure 3-5a is remarkable in several ways. First, the large central hole is formed by seating domain III on top of domains I and IV. This architecture suggests that domain II can be used as a hinge for moving the "lid" formed by domain III away from its seat or "base" formed by domains I and IV. Relative movements between the lid and the base would allow the enzyme, in the shape of a C-clamp, to open or close its jaws. It has been suggested that two regions, one at the two long strands connecting domain II to the base and the other at the β-sheet connecting domains II and III, may serve as hinges for a sideways rotation of domains II and III away from the base.[20] A plausible upward lifting of domain III away from the base has also been suggested.[19]

Second, although the structural domains of proteins generally tend to be formed from contiguous stretches of polypeptides, the *E. coli* Top1 polypeptide folds in a very peculiar way.[15] It first completes domain I and continues into domain IV, but it finishes

17. Hansen, G., et al. 2006. *J. Mol. Biol.* **358:** 1328–1340.

18. Mondragón, A. and DiGate, R.J. 1999. *Structure* **7:** 1373–1383.

19. Rodriguez, A.C. and Stock, D. 2002. *EMBO J.* **21:** 418–426.

20. Feinberg, H., et al. 1999. *Nature Struct. Biol.* **6:** 918–922.

only one-third of domain IV before leaving for domain II. After completing about one-half of domain II, it again temporarily abandons this domain and extends into domain III; only after completing the entire domain III does the polypeptide chain head back to finish folding domains II and IV. This intricate path of the polypeptide is highlighted by the color scheme used in Figure 3-5b: The long (595-amino acid) backbone of the *E. coli* Top1 fragment is represented by a coil whose amino-terminal region is colored dark blue, with the color gradually shifting, along a rainbow-like spectrum of blue, green, yellow, orange, and red, as the polypeptide extends toward its carboxyl terminus. In this depiction, domain IV, with its interwoven segments of different colors, is particularly striking.

Third, tyrosyl residue Tyr-319 (the red amino acid residue labeled Y in Fig. 3-5a), at the active site of the enzyme, is strategically perched on the edge of domain III. There is also a nice cleft in domain IV that could bind the scissile strand. Some movements would be necessary, however, to make room for binding of the DNA strand as well as to properly position, relative to the bound scissile strand, amino acid residues that participate in DNA breakage and rejoining (these include Tyr-319 and several acidic amino acid residues located in domain I, especially Glu-9).[16,21] These conjectures have been supported by several crystal structures of single-stranded DNA oligonucleotides bound to *E. coli* Top1 and Top3.[22]

How Tyr-319 and the other amino acid residues coordinate in their catalysis of DNA breakage or rejoining is not entirely clear. There is strong evidence, however, that the basic mechanisms of DNA breakage and rejoining by the different topoisomerase subfamilies are very similar. In all cases an active-site tyrosine, an arginine adjacent to the tyrosine, and a cluster of several acidic amino acid residues are likely to play significant roles, as we shall see in Chapter 4.

THE MANIPULATION OF DNA STRANDS BY A TYPE IA DNA TOPOISOMERASE

The findings described in the previous sections are consistent with the mechanism outlined in Figure 3-6 for the catalysis of a DNA strand passage by a type IA enzyme.[15] The images labeled a–d in Figure 3-6 represent four major states of the reaction cycle. In Figure 3-6a, the scissile DNA strand is shown to interact mostly with domain IV, but its path must also closely skirt domain III for proper presentation of the DNA strand to the active site tyrosine. Following cleavage of the scissile strand, the lid is shown to move away from the base to open a transient gate in the scissile DNA strand that is sufficiently wide for the passage of another DNA into the interior of the enzyme (Fig.

21. Chen, S.-J. and Wang, J.C. 1998. *J. Biol. Chem.* **273:** 6050–6056; Zhu, C.X., et al. 1998. *J. Biol. Chem.* **273:** 8783–8789.

22. Changela, A., et al. 2001. *Nature* **411:** 1077–1081; Perry, K. and Mondragón, A. 2003. *Structure* **11:** 1349–1358; Changela, A., et al. 2007. *J. Mol. Biol.* **368:** 105–118.

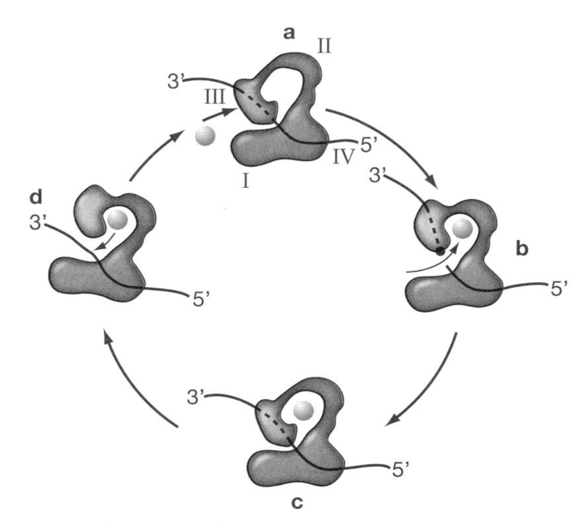

Figure 3-6. The major steps of a proposed reaction cycle of a type IA DNA topoisomerase. See text for explanations. (The motif representing the enzyme is adapted, by permission of Macmillan Publishers Ltd., from Wang, J.C. 2002. *Nat. Rev. Mol. Cell. Biol.* **3:** 430–440.)

3-6b). We shall consider the nature of the passing DNA strand and its interaction with the enzyme in the next section; here it is simply depicted as a circle, as an end-on view. Following entrance of the passing DNA strand, the lid closes to permit the rejoining of the broken scissile strand (Fig. 3-6c). The lid then reopens, this time without breaking the scissile strand, to allow the exit of the DNA segment inside the hole without passing through the scissile strand again (Fig. 3-6d). The reaction cycle ends with closure of the emptied C-clamp to return structure d back to a.

The reaction cycle depicted in Figure 3-6 should be viewed as a conceptual framework, with many of its details yet to be filled in. Nevertheless, it is a good bet that the depiction in Figure 3-6a will prove to be a good resemblance to reality. First, all of the available crystal structures of type IA DNA topoisomerases with bound single-stranded oligonucleotides are consistent with this basic idea. Second, if the enzyme-bridging idea is correct, then there ought to be two different ways for passing a DNA strand into and out of the recess in the interior of the enzyme: one in which the DNA strand goes through an enzyme-operated gate in the DNA scissile strand, and the other without the DNA strand going through the DNA gate. Clearly there would be no net topological change of any kind if the passing DNA strand goes through the opened scissile strand both on its way in and out. Third, each of the steps depicted in the reaction cycle shown can proceed in either direction: from Figure 3-6c to 3-6b or from 3-6b to 3-6c, and so forth. As we have discussed, an enzyme that catalyzes a step A $\rightarrow$ B must also catalyze the reverse step B $\rightarrow$ A. The net direction of a reaction is not determined by the enzyme if it is capable of performing multiple cycles of the same reaction, but rather by the energetics of the overall reaction in each cycle. The relaxation of a negatively supercoiled DNA by *E. coli* Top1, for example, is driven by the higher energy state of the supercoiled DNA relative to its relaxed form (Chapter 2, Footnote 14).

The rate of this reaction, on the other hand, is very much determined by interactions between the enzyme and DNA. In the relaxation of a negatively supercoiled DNA, the rate-limiting step—that is, the bottleneck that determines the overall reaction rate—is probably the formation of the state shown in Figure 3-6a, in which the enzyme binds to a negatively supercoiled DNA. In the case of a positively supercoiled DNA, this step is likely to occur extremely slowly, because the enzyme has a difficult time unwinding the double helix to expose a single-stranded region. Thus, despite the favorable overall energetics for DNA relaxation, the *E. coli* enzyme cannot catalyze this process at an appreciable rate.

The ordering of the reaction steps in the type IA DNA topoisomerase-catalyzed reaction, such as a → b → c → d → a, is rather arbitrary. There is no compelling reason to select that particular order versus, say, a sequence of c → d → a → b → c, or some sequence in reverse, such as c → b → a → d → c. The general principles in enzyme catalysis require only that the states a, b, c, and d not differ greatly in their energies, so that the various enzyme–DNA complexes can coexist. If any of these complexes were in a much lower energy state, it would be very difficult for it to "climb out" of the low-energy sink.

THE NATURE OF THE PASSING DNA STRAND

In topological terms, the relaxation of a negatively supercoiled DNA by a type IA DNA topoisomerase involves the transient cleavage of one DNA strand and the passage of the other DNA strand through the transiently opened gate in the cleaved DNA strand. Thus, statements such as "the type I DNA topoisomerases pass one DNA strand through another" often appear in the literature. In structural terms, however, it remains uncertain whether the passing strand is just a single strand or is a strand in a double-stranded segment, because passing either a single-stranded or a double-stranded DNA segment through a gate in the scissile strand has the same outcome (Fig. 3-7). In Figure 3-7a we see that one strand in a single-stranded bubble in a duplex DNA is cleaved by the topoisomerase, and that the single-stranded region opposite the nick serves as the passing strand. In Figure 3-7b, a duplex DNA segment adjacent to the enzyme-generated break is shown to have crossed the break; upon resealing of the break, the duplex can slip out of the bubble by diffusion (we can better visualize this process by imagining pulling the two duplex ends in the rightmost drawing away from each other).

In concept, the passage of a single-stranded region opposite the enzyme-operated DNA gate in the other strand is simpler (hence the type IA DNA topoisomerases are usually thought to act in this fashion). There are, however, hints that the mechanism

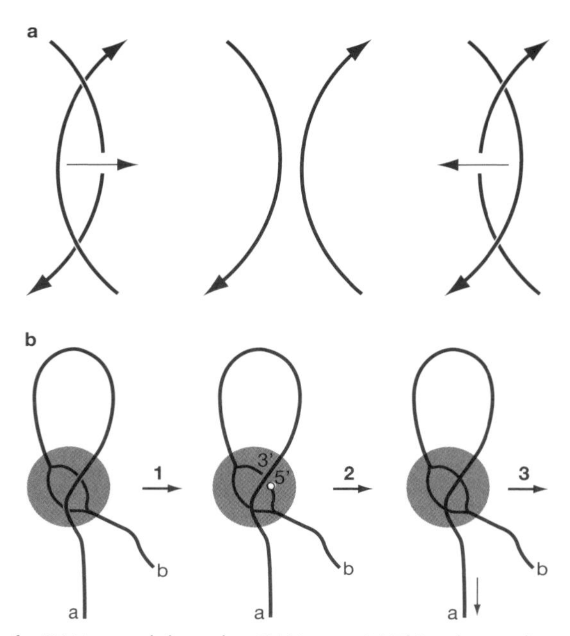

Figure 3-7. Passage of a DNA strand through a DNA gate. (*a*) This scheme shows the passage of one strand of a double-stranded DNA through a transient break in the other strand. The two arcs represent two short segments of DNA strands in a double-stranded DNA ring, the arrow at the tip of each line indicates the 5′ → 3′ direction. The small horizontal arrow marks the path of one strand as it passes through a transient break in the other strand (the scissile strand). Each pair of singly linked strands on the left and right makes two interstrand crossings; the left-side pair contributes a value of –1 to the *Lk* of the DNA ring, and the right-side pair a value of +1 (see Fig. 2-2 for the sign of *Lk*). In either case, the passage of one strand through another eliminates two interstrand crossings and the link between the crossings (*middle*); the strand crossing shown on the *left* therefore increases *Lk* by 1 and that on the *right* decreases *Lk* by 1. (*b*) This scheme shows the possibility of the type IA DNA topoisomerase-mediated passage of a double-stranded DNA segment through a scissile strand in the same DNA molecule. The thick line between a and b represents double-stranded DNA. A type IA enzyme bound to the DNA, represented by the blue sphere, forms a small single-stranded bubble in it (thinner lines). On the *left*, the DNA segment is shown to loop over the enzyme. In the *middle*, the double-stranded segment crosses a strand break formed by the enzyme; on the *right*, the transiently opened DNA gate in the scissile strand is closed after the passage of the duplex DNA segment through it. Diffusion of the duplex DNA out of the single-stranded bubble completes a reaction cycle. See text for details.

could also involve passage of a duplex segment through the transient strand break. First, as we saw earlier, *E. coli* Top1 can catalyze catenation between two double-stranded DNA rings if at least one of the rings contains a nick.[14] The simplest interpretation for this is that the enzyme binds to the DNA strand opposite the nick in the nicked DNA, opens a DNA gate in it, and passes a double-stranded segment of the second DNA ring through this gate before closing the gate. Second, in the four domain structures of *E. coli* Top1 and its closely related enzymes, there is a 2.5-nm hole decorated with positively charged amino acid side chains.[15,17,18,20,22] Model building suggests that this hole is suitable for the binding of a DNA double helix. Third, whereas crystal structures of *E. coli* Top1 or Top3 and their complexes with oligonucleotides have provided strong evidence for the presence of a DNA cleft for binding of the scissile strand, a second binding site for another single strand has not been seen.[15,18,20,22]

On the other hand, in the crystal structure of *A. fulgidus* the four topoisomerase domains enclose a 1.6-nm hole, which is significantly smaller than that observed with the other DNA topoisomerases, and perhaps too small to comfortably accommodate a DNA double helix.[19] The knotting of single-stranded DNA rings by a type IA DNA topoisomerase is also much harder to explain if the passing strand must be in a double-stranded form. It is plausible, however, that the passing DNA can be either single- or double-stranded.

GENOMIC STUDIES REVEAL A UBIQUITOUS ENZYME FOUND IN ALL DOMAINS OF LIFE

The explosive increase in DNA base-sequence information since the late 1970s has fundamentally altered our approach to the identification of RNA or protein products of similar genes in different organisms. When *E. coli* DNA topoisomerase I was first reported in 1971, painstaking biochemical studies were required to establish the presence of similar activities in other bacteria, as well as of a second type IA enzyme in *E. coli*, DNA topoisomerase III (Top3).[23] For many years no type IA DNA topoisomerase activity could be convincingly demonstrated in eukaryotic organisms, other

> A quarter of a century after the debut of the Watson–Crick structure, two powerful DNA sequencing methods were developed for determining the precise arrangements of the four nucleotides in a DNA strand. Another quarter of a century later, the nucleotide sequence of the entire human genome—some three billion base pairs—became known, which provided previously unfathomable details of the human genetic book. The huge nucleotide sequencing capacity of laboratories all around the world has since turned to sequencing the genomes of different organisms across a diverse spectrum, including the genomes of individual human beings for identifying genetic variations among different populations and individuals. The genomic sequences of a large number of organisms, including many representatives of all three domains of life—bacteria, eukarya, and archaea—have already been completely determined, and the total number of completely sequenced genomes continues to increase at a rapid rate.

23. Dean, F., et al. 1983. *Cold Spring Harb. Symp. Quant. Biol.* **47:** 769–777; Srivenugopal, K. S., et al. 1984. *Biochemistry* **23:** 1899–1906.

than in chloroplasts of green plants (believed to be of bacterial origin). Thus for a long time it was thought that the type IA enzymes were present only in prokaryotes—organisms that do not have a membrane-enclosed cell nucleus, and not in eukaryotic organisms, which do.

But in 1989, studies of the gene *EDR1* of the eukaryotic microorganism *Saccharomyces cerevisiae*—baker's yeast—eventually led to its sequencing. The *EDR1* nucleotide sequence, and the deduced amino acid sequence of the polypeptide encoded by it, clearly showed that the gene and its product are closely related to the *E. coli topA* gene and its protein product Top1. The *EDR1* gene was renamed *TOP3* and its product was renamed DNA topoisomerase III (by then DNA topoisomerases I and II had already been identified in the same yeast).[24] Biochemical experiments soon confirmed that the yeast enzyme was indeed a type IA DNA topoisomerase.[25]

The human counterparts of the *E. coli* and yeast type IA enzymes were also initially identified through nucleotide sequencing. In the early 1990s, efforts were made to randomly sequence large numbers of what had been termed "expressed sequence tags," or ESTs—short stretches of nucleotide sequences in genes encoding polypeptides. Around 1996, the search of a large compilation (database) of human EST sequences revealed one member that appeared to encode a stretch of seven amino acid residues found in several type IA DNA topoisomerases of known sequences. This clue led to the sequencing of a longer region containing the common sequence, and the deduced amino acid sequence left very little doubt that it represented a gene encoding a human type IA DNA topoisomerase. Biochemical studies of the protein expressed by the cloned coding sequences of the putative gene soon confirmed this prediction.[26] The human gene was initially termed *TOP3* and its product was designated DNA topoisomerase III. Soon afterward, however, large-scale nucleotide sequencing of an unrelated region of a different chromosome revealed the existence of a different but closely related gene; these structural genes were therefore named *TOP3α* and *TOP3β*, and their products DNA topoisomerases IIIα and IIIβ (Top3α and Top3β), respectively.[27]

The discovery of the first archaeal type IA DNA topoisomerase also had an interesting history. In 1984, biochemical studies had identified an activity from the archaeon *Sulfolobus shibatae*, a microorganism living in sulfurous hot springs, that catalyzes the supercoiling of a DNA ring in the presence of ATP. However, in contrast to the negative supercoiling activity of bacterial gyrase discovered in 1976, which we shall examine in detail in Chapter 5, the archaeal enzyme catalyzes positive supercoiling. Hence the activity was named "reverse gyrase."[28] Nearly a decade would pass

24. Wallis, JW., et al. 1989. *Cell* **58:** 409–419.

25. Kim, R.A. and Wang, J.C. 1992. *J. Biol. Chem.* **267:** 17178–17185.

26. Hanai, R., et al. 1996. *Proc. Natl. Acad. Sci.* **93:** 3653–3657.

27. Kawasaki, K., et al. 1997. *Genome Res.* **7**: 250–261; Seki, T., et al. 1998. *J. Biol. Chem.* **273:** 28553–28556.

28. Kikuchi, A. and Asai, K. 1984. *Nature* **309:** 677–681.

before the enzyme would be classified as a type IA DNA topoisomerase. Nucleotide sequencing of the structural gene encoding the enzyme showed that the carboxy-terminal half of the enzyme has homologies with other type IA DNA topoisomerases, and that the amino-terminal domain of the enzyme contains motifs found in other ATP-utilizing proteins.[29] A type IA DNA topoisomerase not fused to an ATPase domain has also been identified in the hyperthermophile *Sulfolobus sulfataricus*. The archaeal enzyme relaxes negatively supercoiled DNA at 75°C; otherwise its properties resemble those of *E. coli* and yeast Top3.[30]

The accumulation of genomic sequence data provided increasing confidence that sequence comparisons usually afford reliable identifications of homologs in different organisms. And for this reason, in more recent years, biochemical searches of related proteins in different organisms have almost invariably been preceded by sequence searches. In the case of the type IA DNA topoisomerases, scores of genomic sequences have been examined, and there is still only a single exception to the hypothesis that any living organism must possess at least one type IA DNA topoisomerase. This striking ubiquity in all three domains of life, bacteria, eukarya and archaea, is indicative of an ancient origin of the type IA DNA topoisomerases.[31] Does this ubiquity also imply that this subfamily of enzymes plays a key cellular role or roles? The short answer is yes, but we shall postpone more detailed discussions until Chapters 7 and 8, which address the biological roles of all subfamilies of DNA topoisomerases.

29. Confalonieri, F., et al. 1993. *Proc. Natl. Acad. Sci.* **90:** 4753–4757.

30. Dai P, et al. 2003. *J. Bacteriol.* **185:** 5500–5507.

31. See, for example, Forterre, P., et al. 2007. *Biochimie* **89:** 427–446.

CHAPTER 4

A Magic Swivel

"Enzymatic reactions have always had a certain air of magic, perhaps witchcraft. Of course this is due to our imperfect knowledge of what really happens."

John Cornforth

OVER THE CENTURIES, ARTISANS AND ENGINEERS have come up with many ingenious designs of swivels, swivel joints, and swivel connectors. These devices permit two connected parts to rotate relative to each other: A person seated in a swivel chair can spin around relative to the base of the chair, pipes connected by swivel joints can rotate relative to each other, and a fishing swivel allows a fishing line to untwist while it is being cast or retrieved. As we discussed in Chapter 1, one solution to the untangling problem of a replicating DNA would be rotation of the parental double helix about its helical axis. Axial rotation of a lengthy DNA molecule in a crowded and compartmentalized cellular milieu would seem rather difficult, however, if not impossible. Would it not be ideal if swivel joints or connectors could be placed at strategic positions along a long length of DNA, so that it would be necessary only to rotate shorter DNA segments between the swivel joints, rather than a long stretch of DNA extending all the way from a replication fork to the end of a chromosome? Given this, it is not surprising that Nature should come up with its own designs for a swivel or two for DNA.

An enzyme that comes very close to serving as a swivel joint, allowing adjacent DNA segments to rotate relative to each other, was discovered by James J. Champoux and Renato Dulbecco in 1972.[1] This enzyme, detected initially as an activity in extracts of mouse cells that can relax a supercoiled DNA, differs in two important respects from *Escherichia coli* DNA topoisomerase I (known at the time as the ω protein). First,

1. Champoux, J.J. and Dulbecco, R. 1972. *Proc. Natl. Acad. Sci.* **69:** 143–146.

it can relax both positively and negatively supercoiled DNA, and second, it does not require the presence of magnesium ions for its activity. Champoux and Dulbecco called this activity the "DNA untwisting enzyme."[2]

The mouse enzyme would turn out to be the founding member of a new subfamily of DNA topoisomerases now termed the type IB DNA topoisomerases. Activities very similar to that of the mouse enzyme were soon found in many other eukaryotic cells, including those of the fruit fly, sea urchin, calf thymus, and human. These enzymes, typically having a molecular mass of ~90,000 Da or 90 kDa,[3] were assigned the generic name "eukaryotic DNA topoisomerase I" to distinguish them from the type IA bacterial enzymes exemplified by *E. coli* DNA topoisomerase I. A quarter of a century later, however, it would become known that the type IB enzyme is also present in many bacteria—as we shall see in a later section of this chapter.

HALLMARKS OF A TYPE IB ENZYME

Catalysis of DNA strand breakage by all DNA topoisomerases starts with an attack on a DNA phosphorus by an active-site tyrosyl residue (Chapter 2 and Appendix 1); however, the covalent protein–DNA intermediate formed in this transesterification reaction is not always the same. In the case of the type IA DNA topoisomerases, as well as the type IIA and IIB enzymes to be described Chapter 5, DNA strand breakage is accompanied by the covalent attachment of a protein tyrosyl residue to a DNA 5′-phosphoryl group, leaving a 3′-hydroxyl group on the other broken DNA end. By contrast, DNA strand breakage catalyzed by type IB enzymes is accompanied by the covalent attachment of the tyrosyl residue to a DNA 3′-phosphoryl group, leaving a 5′-hydroxyl group on the other broken DNA end.[4]

In both cases, the chemistry is actually rather similar. The linkage of the protein to the DNA strand in the transesterification reaction (whether 5′ or 3′) is determined by the way in which the oxygen atom of the enzyme tyrosyl residue and the phosphate moiety in the backbone of the DNA strand are positioned in the catalytic pocket of the enzyme. It is this positioning, which differs in the type IA and IB enzymes, that determines the final outcome (shown in Fig. 4-1). Figure 4-1a highlights the phosphate moiety in the backbone; here the four oxygen atoms are located at the four corners of a tetrahedron that has the phosphorus atom at its center. The sugar-linked 3′ and

2. Naming a newly discovered enzyme is not as straightforward as it might seem. The original name "DNA untwisting enzyme" for a topoisomerase was problematic because the extent of coiling between the two strands of the DNA actually increases when a negatively supercoiled DNA, which is underwound, is relaxed by the topoisomerase.

3. Masses of atoms and molecules are expressed in units of Daltons (Da) (in honor of John Dalton, 1766–1844, British chemist who developed the atomic theory of matter); an ^{16}O oxygen atom has a mass of 16 Da, and the total mass of 1 mole or 6.023 x 10^{23} atoms of ^{16}O is 16 g.

4. Champoux, J.J. 1977. *Proc. Natl. Acad. Sci.* **74:** 3800–3804.

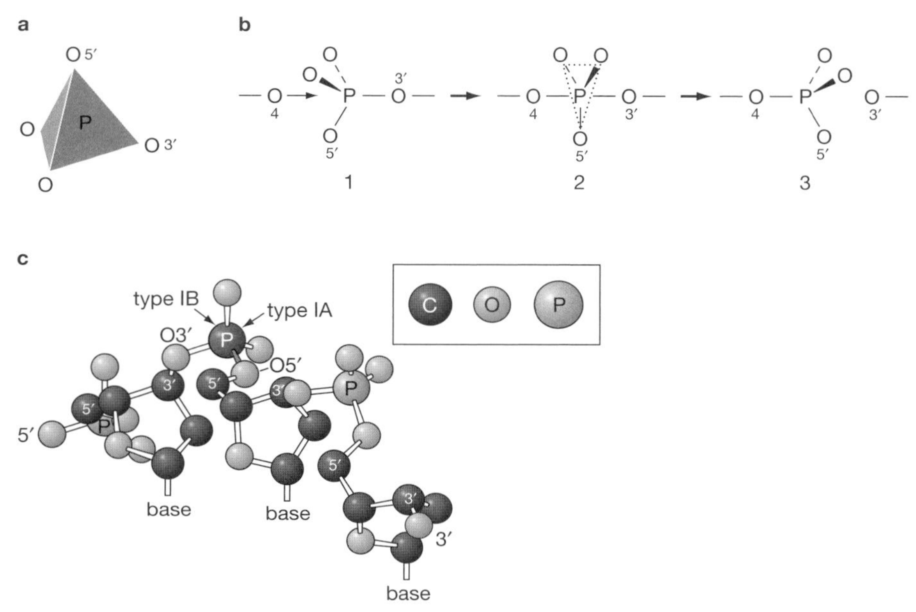

Figure 4-1. The type IA and IB DNA topoisomerases form different covalent intermediates in their breakage of a DNA strand because of the differences in their positioning of the reacting groups. (*a*) The geometry of a phosphate group in the DNA backbone. (*b*) The reaction sequence represents the "in-line phosphoryltransfer mechanism." O4 is the oxygen extending from the fourth carbon of the phenyl ring of the active-site tyrosine of a type IA DNA topoisomerase. The solid wedge and dotted line between the two non-bridging oxygens and the central phosphorus represent P–O bonds that are pointing outward and inward, respectively, from the plane of the paper. (*c*) The chemical structure of part of the DNA backbone shows points of attack by Type IA and IB topoisomerases. The carbon atoms are represented by black spheres (with C5′ and C3′ labeled), the oxygen by gray spheres (with O5′ and O3′ labeled), and the phosphorus by blue spheres. See text for details. The hydrogen atoms are omitted for clarity. See text for details.

5′ oxygen atoms are a part of the DNA strand backbone, and the other two are "non-bridging oxygens." Figure 4-1b illustrates the "in-line phosphoryl transfer" mechanism, often observed in reactions involving phosphorus-containing compounds. On the left (Fig. 4-1b1), an oxygen atom of a type IA DNA topoisomerase (O4), is shown to approach a DNA phosphorus to initiate transesterification between the topoisomerase and the DNA strand. The middle structure (Fig. 4-1b2) depicts the formation of a reaction intermediate: The phosphorus atom is bonded to five oxygen atoms, with a new (fifth) bond between the attacking O4 and the phosphorus. This intermediate is in the shape of a bipyramid: the O4 and O3′ oxygen atoms are located at the apices of the bipyramid, and the P atom is at the center of a triangle formed by O5′ and the two non-bridging oxygens. On the right (Fig. 4-1b3), the P–O3′ bond is broken and the P atom and its four associated oxygen atoms resume a tetrahedral geometry. The mechanism depicted is called "in-line phosphoryl transfer" because the attacking group (O4) and the leaving group (O3′) in this reaction lie in a straight line along the apices of the bipyramid-shaped intermediate (Fig. 4-1b2). In Figure 4-1c, three nucleotides along a DNA strand are shown according to their geometry in a Watson–Crick DNA double helix, and the alternative points of attack on a phosphorus atom by the two types of topoisomerase enzyme are marked. As indicated in Figure 4-1c, the O4 atom of a type IA enzyme attacks the scissile phosphorus from the northeast direction, and from the opposite side of the P—O3′ bond of the phosphate. According to the in-line mechanism shown in Figure 4-1b, covalent bond formation between O4 and P would join the enzyme tyrosyl group to a 5′-phosphoryl group in the DNA, leaving a 3′-hydroxyl on the other end of the broken DNA strand. Similarly, in the case of a type IB enzyme, the O4 atom would approach from the northwest direction shown in the figure, joining the active-site tyrosyl group to a 3′-phosphoryl group in a DNA strand, leaving a 5′-hydroxyl at the other end of the DNA strand.

Whereas both type IA and IB DNA topoisomerases break one DNA strand at a time (hence their classification as type I), and both act as monomers in their catalysis of DNA strand breakage and rejoining, the two subfamilies of proteins share very little similarity in their amino acid sequences. The structural features of the type IB enzymes are also very different from those described in Chapter 3 for the type IA enzymes, as demonstrated below in crystallographic studies of human DNA topoisomerase I (Top1).

Human Top1 is a polypeptide of 765 amino acid residues with a molecular mass of 91 kDa. Several crystal structures of covalent and noncovalent complexes between the human enzyme and DNA have been known since 1998, and these structures have provided much insight as to how a type IB enzyme performs its magic.[5] In studying two of these structures, a 64-kDa enzyme reconstituted from two protein fragments was used: A large fragment contained residues 175–659, and a small one containing residues 713–765. These two human Top1 fragments had been shown to form a fully active 1:1 complex, demonstrating that the regions absent in the reconstituted 64-kDa fragment are dispensable for the enzymatic activity of intact human Top1. In the third crystal, a 70-kDa human Top1 fragment spanning amino acid residues 175–765 was used. For noncovalent complex formation, both the 64- and the 70-kDa enzyme fragment contained an Y723F mutation, in which a phenylalanine (F) was substituted for the active-site tyrosine Y723. Whereas this Y723F substitution has minimal effect in terms of DNA binding, the absence of the critical tyrosyl OH group in phenylalanine makes the mutant enzyme incapable of DNA cleavage, and thus the mutant enzyme fragments are ideally suited for studies of noncovalent complex formation.

The structures of the two noncovalent complexes are essentially identical in regions where the 70- and 64-kDa polypeptides overlap. Figure 4-2 presents two views

5. Redino, M.R., et al. 1998. *Science* **279:** 1504–1513; Stewart, L., et al. 1998. *Science* **279:** 1534–1541; Staker, B.L., et al. 2002. *Proc. Natl. Acad. Sci.* **99:** 15387–15392.

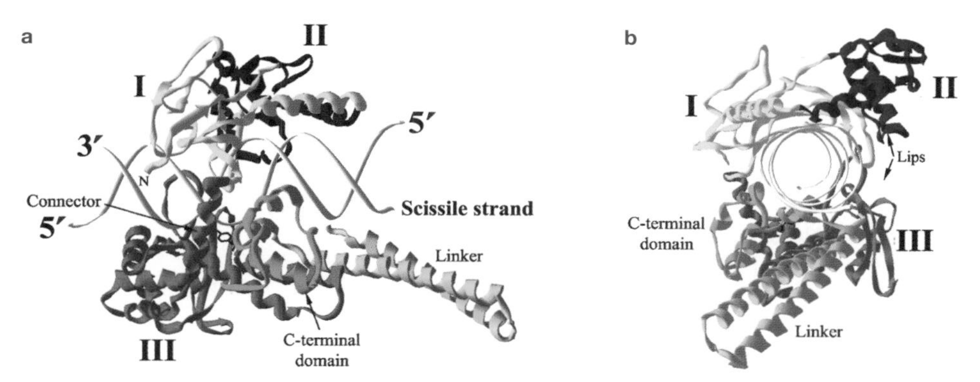

Figure 4-2. Two views of a noncovalent complex between DNA and a 70-kDa fragment spanning amino acid residues 175–765 of human DNA topoisomerase I. The Roman numerals I, II, and III denote the subdomains of the protein core, colored respectively in yellow, blue, and red; the letter N denotes the amino-terminus of the visible portion of the polypeptide in the crystal. The small carboxy-terminal domain is colored in green, and a pair of long α-helices form a linker that connects this domain to subdomain III. The Phe-723 residue of the Y723F mutant enzyme used in forming the crystal is shown in black at the left edge of the green carboxy-terminal domain (note that Phe-723 in the mutant replaces the active site Tyr-723 in the normal or wild-type protein). (Reprinted, with permission, from Champoux, J.J. 2001. *Annu. Rev. Biochem.* **70:** 369–413.)

of the structure of the noncovalent complex of the 70-kDa fragment. The protein can be viewed as a large "core domain" connected to a small carboxy-terminal domain through a "linker" composed of a pair of long α-helices. The core domain is further subdivided into regions labeled I, II, and III. The region between amino acid residues 175 and 214, as well as a six-amino- acid stretch joining the carboxyl terminus of III to the linker region, is not sufficiently well oriented in the crystal to be visible by X-ray diffraction. The 22-bp DNA segment in the noncovalent complex (shown here as a double helix) was found to be present in the B-helical form in the crystal structures. In the crystal structure, Phe-723 (black-outlined hexagonal shape in Fig. 4-2a) is close to the backbone phosphorus at which cleavage of the scissile DNA strand by human Top1 would be expected (without the Y723F substitution).

The subdomains I and II form a "cap" that connects to the "base" of the protein, formed by subdomain III and the carboxy-terminal domain, through a long α-helix in subdomain III (labeled as the "connector" in Fig. 4-2a). As can be seen in the projection of the human Top1–DNA complex shown in Figure 4-2b, the cap and base of the 70-kDa protein encircle the DNA double helix through interactions between a pair of loops dubbed the "lips," one in subdomain II and the other in subdomain III. For an unbound enzyme to circle around a DNA double helix, it must be able to assume a more open conformation than the DNA-bound form depicted in the figure; presumably, this conformational change involves breaking the lip contacts and moving the cap upward from the base.

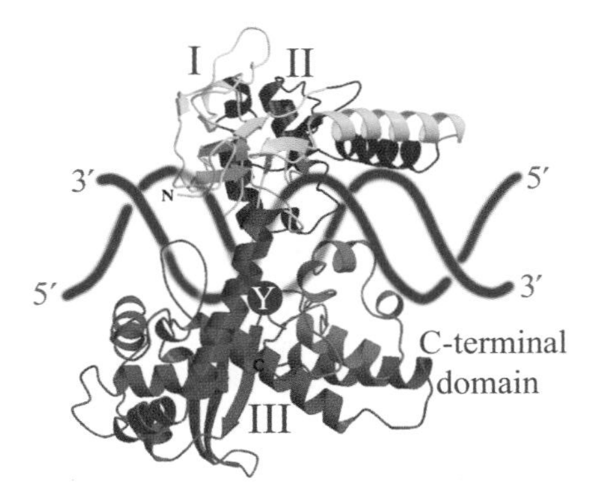

Figure 4-3. Illustration of the structure of a covalent complex between a 22-bp DNA fragment and a reconstituted 64-kDa human DNA topoisomerase I composed of amino acid residues 175–659 and 713–765. The structure is shown as an overlay of the DNA backbones and the protein. The blue sphere (labeled "Y") represents the active-site tyrosyl residue Tyr-723, which is covalently linked to the 3′-end of the cleaved scissile strand. (Drawings are based on data from Redino, M.R. et al. 1998. *Science* **279:** 1504–1513.)

Figure 4-3 depicts the covalent complex between the 64-kDa reconstituted human Top1 fragment and the 22-bp DNA (the same fragment shown in Fig. 4-2). To enhance formation of this covalent complex, the anticipated scissile P—O bond in the DNA was first replaced by a P—S bond. Replacing the oxygen with a chemically similar sulfur did not prevent the cleavage of the DNA strand by the enzyme, and the usual covalent complex, with the enzyme attached to a DNA 3′-phosphoryl group, was formed. The formation of the covalent complex in this particular case, however, left a sulfhydryl (—SH) rather than a hydroxyl (—OH) group at the 5′ end of the broken DNA strand. The enzyme-mediated rejoining of the broken DNA strand in this covalent adduct was much slower than the corresponding reaction with a hydroxyl group at the 5′ end of the broken DNA strand. The reason for this decrease in reaction rate is not clear. One plausible contributing factor is that sulfur, being less electron-rich than oxygen, is less proficient in attacking the electron-poor phosphorus to form a covalent bond. Whatever the reason, this reduction in the rate of rejoining of the broken DNA strand led to accumulation of the covalent adduct.

The positions of the various protein domains in the covalent complex vary but little from their counterparts in the noncovalent complexes, and the largest differences are observed in the immediate vicinity of the catalytic pocket for DNA strand cleavage. Similarly, there is little structural difference in the enzyme-bound DNA in the noncovalent and covalent complexes: The 22-bp DNA exhibits the B-form double-helix structure, and minor structural differences among the three crystals are seen mostly in the immediate vicinity of the site of DNA strand breakage.

A MAGIC SWIVEL

At first glance, the relative movements between the "cap" and the "base" of human DNA topoisomerase I during its binding to DNA may seem similar to movements between the "lid" and the "base" of a type IA enzyme in its manipulation of DNA, as we saw in Figure 3-7. Beyond this superficial resemblance, however, the type IB and type IA enzymes are structurally very different. The presence of the DNA in a B-helix form in the human Top1–DNA complexes, either before or after the cleavage of the scissile DNA strand, further confirms earlier biochemical results that a type IB enzyme acts on a double-stranded DNA.[6] In sharp contrast to a type IA enzyme, there is no indication that strand cleavage by a type IB enzyme requires that the scissile strand be in a single-stranded form.

The retention of the double-helical geometry of DNA in the case of a type IB topoisomerase enzyme suggests that this enzyme functions as a simple swivel in a DNA double helix (e.g., after strand cleavage, the DNA segment downstream from the cleavage site can rotate axially, relative to the covalently anchored upstream DNA segment, around one of the several single bonds in the intact strand directly opposite the nick [see Fig. 1-7 in Chapter 1]). The foregoing model for a type IB DNA topoisomerase, termed the "DNA rotation" model, is also consistent with the kinds of DNA–enzyme interactions that can be seen in the crystal structures.[5,7] In these structures, protein–DNA interactions occur mainly within a stretch of about a dozen base pairs in the central part of the 22-bp DNA used in complex formation, and the enzyme interacts more closely with base pairs immediately upstream of the cleavage site than with those downstream from it. Here, upstream and downstream refer to the 5′→3′ direction in the scissile DNA strand. By convention, the nucleotides in the scissile strand are enumerated by their positions relative to the site of cleavage, which occurs between nucleotides –1 and +1; it is the 3′ end of the –1 nucleotide that becomes covalently linked to the type IB enzyme. Nucleotides downstream from the cleavage site are denoted by positive numbers, +1, +2, +3, whereas those upstream of the cleavage site are denoted by negative numbers, –3, –2, and –1. In all covalent and noncovalent human Top1–DNA complexes, nearly three-quarters of the close contacts between the enzyme and DNA involve base pairs –5 through –1. Furthermore, protein–DNA interaction downstream of the cleavage site appears to involve primarily a ring of positively charged residues in the enzyme. These findings suggest that the region of the enzyme downstream from the cleavage site could serve as a sheath lined with positive charges, inside which the negatively charged DNA segment containing the 5′-hydroxyl end could rotate without excessive hindrance.[5]

From this discussion we may conclude that Nature has come up with a magical swivel: The swivel can be put into DNA, or removed from it, whenever and wherever a swivel is needed. The advantage of such a mobile swivel is that the DNA untanglement problem can be alleviated with only transient disruptions of DNA strand continuity. Whereas both type IA and IB DNA topoisomerases act by transiently nicking a

6. See, for example, Been, M.D., and Champoux, J.J. 1984. *J. Mol. Biol.* **180:** 515–531.

7. The model is often termed the "controlled rotation model," but the mechanistic meaning of the adjective "controlled" is unclear.

DNA strand, it is the type IB enzymes that come close to the idea of a simple swivel: They allow two adjacent DNA segments to rotate longitudinally relative to each other, and, at least in principle, many revolutions can occur between the enzyme-mediated cleavage of a DNA strand and the subsequent rejoining of that strand (this issue will be discussed further in Chapter 6). By contrast, in the enzyme-bridging model described in Chapter 3 for the type IA DNA topoisomerases, multiple rotations between successive strand-breakage and -rejoining events are not possible.

DWARFS IN THE IB FAMILY

In 1977, William Bauer and his associates detected an activity similar to eukaryotic DNA topoisomerase I in extracts of vaccinia virus cores.[8] Vaccinia virus is a member of the poxvirus family, of which the smallpox virus, variola, is the best known and most feared member. The viral core-associated activity that Bauer and his colleagues detected seemed to differ from the archetype eukaryotic DNA topoisomerase I in several subtle ways, which suggested that it might be an enzyme encoded by the poxvirus, rather than a host cell activity encapsulated in the virus particles.

A decade elapsed before it was finally established that the virus-associated activity was indeed encoded by the virus, and that it is a type IB DNA topoisomerase of a size much smaller than the type IB enzyme of the host cells.[9] In their 1987 work, Stewart Shuman and Bernard Moss applied reverse genetics to identify the gene encoding the vaccinia virus enzyme. They first purified the putative viral enzyme to a high degree of purity, and then determined a sequence of 22 amino acid residues at its amino terminus. By the late 1980s, DNA sequencing was already in full bloom, and the nucleotide sequences of many regions of vaccinia virus DNA had already been determined. By scanning the available sequences of the viral DNA, Shuman and Moss were able to identify a gene encoding a protein that had the exact sequence of the 22 amino-terminal residues they determined. This viral gene was deduced to encode a protein of 314 amino acids, with a calculated molecular mass of 36,700 Da (a little more than one-third that of an archetypical eukaryotic DNA topoisomerase I). When the amino acid sequence of the viral enzyme was compared with that of yeast DNA topoisomerase I, it was clear that in spite of their large size difference, the two enzymes are closely related. Alignment of the viral and yeast enzyme sequences revealed the presence of a homologous 117-amino-acid stretch in them, which spans residues 116–228 of the viral enzyme and residues 406–522 of the yeast enzyme (Fig. 4-4).[9]

Other poxviruses were soon shown to encode enzymes very similar to the vaccinia virus topoisomerase. These viral enzymes all fall within a narrow size range and

8. Bauer, W.R., et al. 1977. *Proc. Natl. Acad. Sci.* **74:** 1841–1845.

9. Shuman, S. and Moss, B. 1987. *Proc. Natl. Acad. Sci.* **84:** 7478–7482.

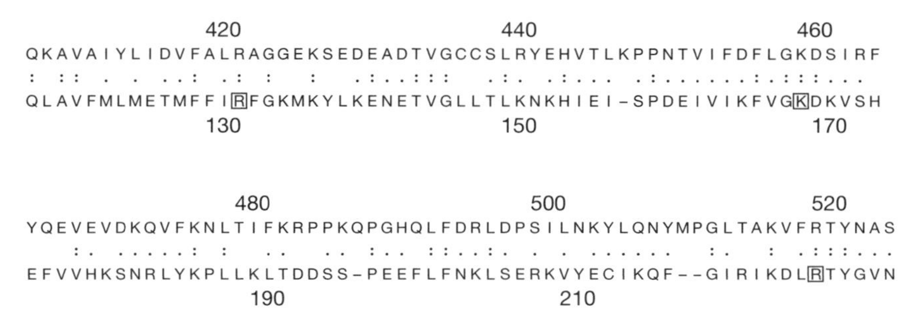

Figure 4-4. Alignment of the amino acid sequence of vaccinia topoisomerase in the region spanning rsidues 116–228 (*lower* sequence) and that of *Saccharomyces cerevisiae* Top1 in the region spanning residues 406–522 (*upper* sequence). The amino acids are shown in one-letter codes. A high degree of sequence homology between the two enzymes is evident. Amino acids at corresponding positions in the two polypeptides that are identical are marked by two dots between the sequences, and those that are not identical but have chemically similar side chains are marked by a single dot between them. Several other regions of significant amino acid sequence homology, including those around the two other members of the catalytically important pentad (H265 and Y274 of the vaccinia topoisomerase and their counterparts in the yeast enzyme, H and Y being the one-letter codes for the amino acids histidine and tyrosine, respectively), are not shown here (see, e.g., Caron, P.R. 1999. *Methods Mol. Biol.* **94:** 279–316). The boxed amino acids in the vaccinia enzyme sequence, R130, K167, and R223 (R and K being the one-letter codes for the amino acids arginine and lysine, respectively), are three of the highly conserved pentad in all type IB enzymes; identical amino acids are present at corresponding positions in *S. cerevisiae* topoisomerase 1. (Redrawn, with permission, from Shuman, S. and Moss, B. 1987. *Proc. Natl. Acad. Sci.* **84:** 7478–7482.)

exhibit a high degree of amino acid sequence homology among them. For example, the amino acid sequence of the enzyme of the dreaded variola virus differs at only three positions from that of the vaccinia enzyme: The amino acids asparagine, glycine, and lysine at positions 24, 47, and 159 in the vaccinia enzyme are, respectively, an aspartic acid, glutamic acid, and aspartic acid in the variola enzyme.

Fifteen years after the identification of the gene encoding poxvirus topoisomerase, DNA sequencing would again show that proteins very similar to the poxvirus enzymes are also encoded by genes of a diverse genera of bacteria, including the human pathogens *Mycobacterium avium* and *Pseudomonas aeroginosa*.[10] Biochemical studies of the predicted protein product of such a gene in *Deinococcus radiodurans* (a bacterium dubbed the toughest bacterium in the *Guinness Book of World Records* because of its extremely high resistance to radiation) soon confirmed that the bacterial gene indeed encodes a type IB DNA topoisomerase. These results thus demonstrated that enzymes of the class "eukaryotic DNA topoisomerase I" are present in both eukaryotes and prokaryotes. Together with the poxvirus topoisomerases, they form the type IB subfamily of enzymes.

10. Krogh, B.O. and Shuman, S. 2002. *Proc. Natl. Acad. Sci.* **99:** 1853–1858.

It is uncertain whether members of the type IB family also exist in archaea, which, together with bacteria and eukarea, form the three domains of life. In 1993, an enzyme with biochemical properties similar to those of the type IB DNA topoisomerases was found in *Methanopyrus kandleri*, a methanogen that grows at temperatures as high as 110°C.[11] The amino acid sequence as well as the three-dimensional structure of a 61 kDa amino-terminal fragment of the *M. kandleri* enzyme shows, however, few signatures of the other type IB enzymes.[12] Thus, enzymes such as the *M. kandleri* "DNA topoisomerase V" are probably very distant relations, rather than certified members, of the DNA topoisomerase family.

SMALL BUT BEAUTIFUL

Despite their small size, the type IB enzymes encoded by poxviruses and bacteria such as *D. radiodurans* are fully capable of relaxing negatively or positively supercoiled DNAs. Among these dwarfs of the type IB family, the vaccinia enzyme is the one that has been most extensively studied.[13] The enzyme consists of two domains, an 80-amino-acid amino-terminal domain and a larger carboxy-terminal domain containing the rest of the protein. In contrast to the much larger type IB DNA topoisomerases like human and yeast DNA topoisomerase I, which exhibit only a very weak preference for certain DNA sequences, the petit viral enzymes require a double-stranded DNA substrate that contains the pentameric sequence 5′-CCCTT-3′, in the scissile strand. The enzyme transiently cleaves the strand after the last T in the pentameric sequence, and is largely inactive in the absence of this recognition sequence.

Biochemical evidence suggests that the small amino-terminal domain and larger carboxy-terminal domain of vaccinia topoisomerase are joined at a flexible hinge, and that together the two domains embrace a DNA double helix at a 5′-CCCTT-3′ site. The two domains were first crystallized separately and their structures determined by X-ray diffraction.[14,15] These structures, together with the results of extensive biochemical studies, suggested a detailed three-dimensional model of a DNA-bound poxvirus topoisomerase.[14] Subsequently, two crystal structures of DNA-bound variola topoisomerase, one with the intact enzyme covalently linked to one strand of a short DNA duplex containing the pentameric recognition sequence, and the other with the enzyme noncovalently bound, were determined.[16] In resemblance to the case for human Top1,[5] the structures of the covalent and noncovalent variola topoisomerase–DNA complexes are

11. Slesarev, A.I., et al. 1993. *Nature* **364:** 735–737.
12. Belova, G.I., et al. 2001. *Proc. Natl. Acad. Sci.* **98:** 6015–6020; Taneja, B., et al. 2006. *EMBO J.* **25:** 398–408.
13. Reviewed in Shuman, S. 1998. *Biochim. Biophys. Acta.* **1400:** 321–337.
14. Sarma, A., et al. 1994. *Structure* **2:** 767–777.
15. Cheng, C., et al. 1998. *Cell* **92:** 841–850.
16. Perry, K., et al. 2006. *Mol. Cell* **23:** 343–354.

remarkably similar, and significant differences are found only in the vicinity of the active site. Furthermore, key structural features of the DNA-bound poxvirus enzyme, revealed by the earlier model deduced from biochemical data and the crystal structures of the two separate domains of vaccinia topoisomerase, are largely borne out by the directly determined structures of DNA-bound variola enzyme.

The structure of the covalent variola enzyme–DNA complex is depicted in Figure 4-5.[16] A "suicide" DNA substrate (whose sequence is shown in the top of Fig. 4-5a) is used in forming this complex. The pentameric recognition sequence 5′-CCCTT-3′ of the poxvirus enzyme is present in the scissile strand (marked by the arrow). Upon formation of the covalent intermediate, the 3′-end of the scissile strand becomes covalently linked to the active-site tyrosyl residue of the enzyme, but the other end, a 5′-ATT-3′ trinucleotide, diffuses away from the enzyme–DNA complex. Rejoining of this trinucleotide to the covalent intermediate is rather inefficient, owing to the very low concentrations of both, and so the covalent complex accumulates in the reaction mixture.[17]

As shown in Figure 4-5b, the poxvirus enzyme, like human Top1, also encircles the DNA. But instead of a pair of β strands that closes the "lips" of the human enzyme, it is the formation of an ionic bond between two oppositely charged amino acid side chains, one in the β5 strand of the small amino-terminal domain and the other in the α5 helix in the large carboxy-terminal domain, that completes the enclosure of the DNA in the C-clamp-shaped protein. The carboxy-terminal domain of the viral enzyme contains two parts that are structurally similar to the two catalytically important domains in human DNA topoisomerase I: human Top1 core subdomain III and a 19-amino-acid stretch wherein the active-site tyrosine resides. Thus it appears that miniaturized type IB enzymes, exemplified by the vaccinia and variola enzymes, have retained all the essential functional units of their much larger kin, the "eukaryotic" DNA topoisomerase I, but have shed all the "bells and whistles."

The structures of poxvirus topoisomerase–DNA complexes show that there are many interactions between amino acid residues of the enzyme and the DNA bases.[15,16] Four amino acid side chains in the amino-terminal domain of the enzyme, Asn-69, Tyr-70, Tyr-72, and Arg-80, and three amino acid side chains in the carboxy-terminal domain, Lys-133, Tyr-136, and Lys-167, are in close contact with the core recognition sequence; three additional amino acid side chains, Lys-135, Arg-206, and Tyr-209 contact base pairs upstream of the core recognition sequence. No corresponding contacts are seen in the human Top1–DNA complexes, and the presence of such contacts only in the viral enzyme explains its sequence specificity. Surprisingly, even though four amino acid residues in the amino-terminal domain of the poxvirus enzyme (Asn-69,

17. Strictly speaking, a suicide substrate in an enzyme reaction is one that inactivates the enzyme as it is being converted to the reaction product(s). Here the poxvirus enzyme remains fully capable of rejoining the enzyme-attached DNA to an appropriate partner, and it is inactivated only in the sense that the rejoining of the two broken ends of the DNA is extremely slow at low concentrations of the enzyme-attached DNA and of DNA fragments bearing an appropriate 5′-OH end.

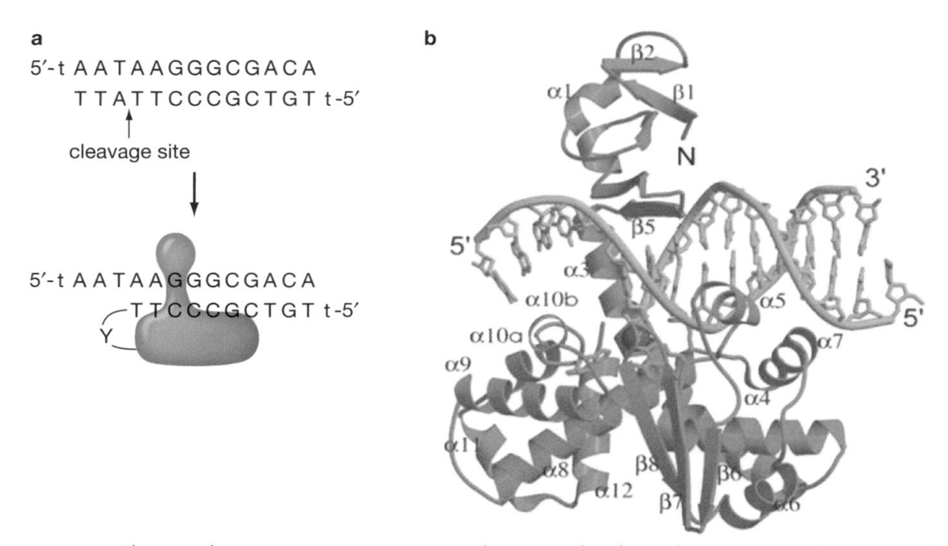

Figure 4-5. The variola topoisomerase. (*a*) A covalent complex forms between a 13-bp DNA substrate and variola topoisomerase. Each strand of the DNA had an extra T at its 5′ end; the 5′-CCCTT-3′ sequence, recognized by the poxvirus topoisomerases, is present near one end of the DNA fragment. Cleavage of the DNA by the bipartite topoisomerase (shaded in blue in the lower panel) leads to covalent joining of the –1 T to the enzyme through a tyrosyl residue (Y), and to loss of the trinucleotide fragment at the 3′ end of the scissile strand. (*b*) A crystal structure of the variola topoisomerase covalent complex as depicted at the *bottom* of a. See the text for details. (Reprinted, with permission of Elsevier, from Perry, K., et al. 2006. *Mol. Cell* **23:** 343–354.)

Tyr-70, Tyr-72, and Arg-80) appear to be involved in recognition of the core sequence 5′-CCCTT-3′, deleting the 80-amino-acid amino-terminal domain of the vaccinia virus enzyme only diminishes its affinity for DNA, but neither abolishes its supercoil removal activity nor eliminates its specificity for the core recognition sequence.[18] The truncated vaccinia enzyme, spanning amino acid residues 81–314, probably represents the simplest design of a DNA swivel.

The small size of poxvirus topoisomerase made it particularly amenable to detailed mutational studies, whereby individual amino acid residues in the enzyme are altered and the mutant enzymes characterized.[13] These studies indicated that in the catalysis of DNA breakage and rejoining, a quartet consisting of Arg-130, Lys-167, Arg-223, and His-265 is of key importance in addition to the active-site tyrosine Tyr-274. Significantly, amino acid residues corresponding to this quartet are also found in the larger type IB enzymes like human and yeast Top1. On the basis of sequence alignment as well as structural and mutational analysis, this quartet is identified to constitute Arg-488, Lys-

18. Cheng, C. and Shuman, S. 1998. *J. Biol. Chem.* **273:** 11589–11595.

532, Lys-590, and His-632 in human DNA topoisomerase I (as noted before, the active-site tyrosine in the human enzyme is Tyr-723). A similar quartet is found also in the bacterial type IB enzyme, though the position corresponding to that of His-265 in the vaccinia enzyme is occupied by an asparagine (Asn-280 in the *D. radiodurans* enzyme).[11] The plausible roles of the highly conserved residues are described in Appendix 2.

AN EQUID OF A DIFFERENT STRIPE

A striking finding in studies of the three-dimensional structures of the type IB DNA topoisomerases is that they are more closely related to a class of enzymes termed the "tyrosine site-specific recombinases" than to the other subfamilies of DNA topoisomerases.[5,15] Site-specific recombinases promote recombination between a pair of DNA sites having particular nucleotide sequences. These enzymes are subdivided into the tyrosine and serine types depending on whether their catalysis of transesterification utilizes an oxygen on a tyrosine or on a serine residue in initiating DNA strand breakage. Among the tyrosine site-specific recombinases that utilize a tyrosine as the shock trooper, phage (bacteriovirus) λ integrase, product of the phage λ *int* gene, is the earliest known and one of the most extensively studied.[19] The reaction catalyzed by this enzyme is illustrated in Figure 4-6. Here the notation BOB′ denotes a site composed of a string of three sequence elements B, O, and B′ at a particular location on the bacterial DNA, and the notation POP′ denotes a site on the phage DNA composed of three sequence blocks P, O, and P′. The pair of sites share a common sequence block O, a 15-base-pair stretch with a sequence 5′-GCTTT*TTTATACTAA-3′ on one strand, and the complementary sequence 3′-CGAAAAAATAT*GATT-5′ on the other strand (the asterisks in these sequences indicate the positions of the recombinase-mediated DNA strand breakage and rejoining). The recombinase cuts and rejoins a total of four DNA strands within the common core sequences in BOB′ and POP′. Its action differs from that of a topoisomerase, however, in that the two broken halves of each DNA are rejoined not to each other, but to the broken halves of the other DNA, to yield BOP′ and POB′. The result of this reaction, when POP′ is present in a ring-shaped phage DNA, is the insertion or "integration" of the phage DNA into the bacterial chromosome, and hence the enzyme promoting this reaction is termed an integrase (Int).[20]

---

19. Mizuuchi, K. and Nash, H.A. 1976. *Proc. Natl. Acad. Sci.* **73:** 3524–3528; Kikuchi, Y. and Nash, H.A. 1979. *Proc. Natl. Acad. Sci.* **76:** 3760–3764; Craig, N.L. and Nash, H.A. 1983. *Cell* **35:** 795–803. For a review of the early studies of this enzyme, see Landy, A. 1989. *Annu. Rev. Biochem.* **58:** 913–949.

20. The reaction depicted in Figure 4-6 may give the impression that a site-specific recombinase like the λ integrase makes two double-stranded cuts in BOB′ and POP′ and then switches the four broken halves in the rejoining step to give the final products. But results obtained largely in the 1980s support a mechanism involving two sequential events, in which one DNA strand in each of the pair of DNA duplexes is broken at each step and rejoined to the other broken strand. See Nash, H. 1987. In Escherichia coli *and* Salmonella*: Cellular and molecular biology* (eds. F.C. Neidhardt, et al.), pp. 2363–2376. ASM Press, Washington, D.C.

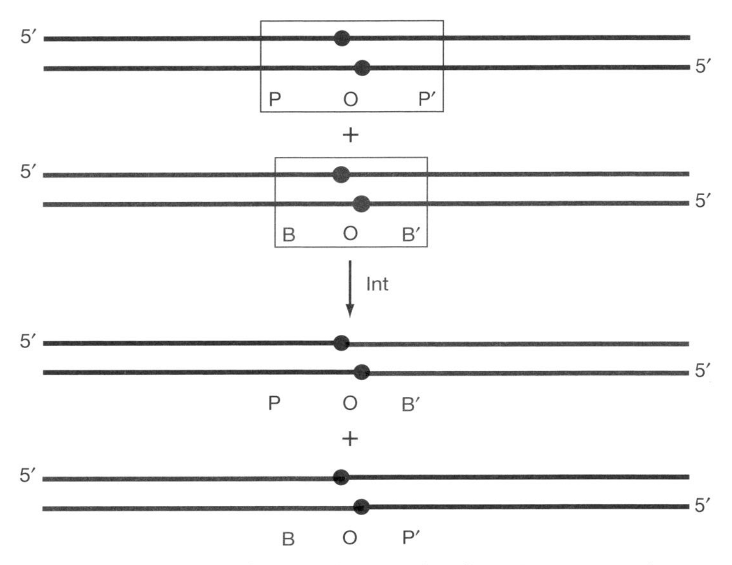

Figure 4-6. The catalysis of site-specific recombination by phage λ integrase. The sequence represented by BOB′ is a block of elements in the bacterial genome; POP′ is the corresponding region in the bacteriophage λ genome. Int is the λ integrase, which promotes the recombination event between the two sequence blocks. See the text for details.

Biochemical studies in the late 1970s and early 1980s, pioneered by Howard Nash and his colleagues, revealed that under certain conditions λ integrase itself can catalyze the relaxation of either positively or negatively supercoiled DNA. (Typically site-specific recombination by λ integrase requires participation of an *E. coli* architectural protein called IHF, for "integration host factor.") Furthermore, the λ integrase-catalyzed supercoil removal reactions require no magnesium (II) ions, and were found to involve a covalent protein–DNA intermediate in which the enzyme is attached to a phosphoryl group at the 3′ end of the broken DNA strand.[19] These properties of the λ integrase were reminiscent of the characteristics of the type IB DNA topoisomerases, and provided the earliest hint that phage λ integrase, and other similar recombinases, might be related to the type IB DNA topoisomerases.

By the time the three-dimensional structures of human DNA topoisomerase I and vaccinia topoisomerase were determined in the late 1990s,[5,15] the three-dimensional structures of several site-specific recombinases had also been solved.[21] A comparison of the structures of the two classes of enzymes indicated that they resemble each other in their catalytic domains. There is extensive structural similarity between domain III

21. Reviewed in Gopaul, D.N. and van Duyne, G.D. 1999. *Curr. Opin. Struct. Biol.* **9:** 14–20; Chen, Y. and Rice P.A. 2003. *Annu. Rev. Biophys. Biomol. Struct.* **32:** 135–159.

of human DNA topoisomerase I and a region in a tyrosine recombinase HP1.[5] Furthermore, residues corresponding to the catalytically important quartet (Lys-167, Arg-130, Arg-223, and His-265 in the vaccinia topoisomerase, and Arg-488, Lys-532, Lys-590, and His-632 in human Top1) are also present in various tyrosine recombinases.[5,15] Together, the biochemical and structural similarities between the type IB DNA topoisomerases and tyrosine recombinases strongly suggest that the two classes of enzyme are evolutionarily related.[5,15]

It thus appears that the two subfamilies of type I DNA topoisomerases, type IA and type IB, have rather different lineages: Nature has apparently invented the type I DNA topoisomerases twice! The type IB enzyme and the tyrosine recombinases presumably originated from a common catalytic core, and additional domains were subsequently acquired to modulate interactions between these proteins and DNA, as well as interactions among different parts of the enzymes. Nature seems innovative on one hand yet conservative on the other, and it tends to stick with successful inventions. Intricate machineries invented by Nature and by humans thus share a feature: Among the human inventions, the myriads of complex machines share many common elements like gears, wheels, connecting bars and motors, and swivels; among Nature's inventions, seemingly different proteins often share common domains and building blocks.

CHAPTER 5

Moving One DNA Double Helix through Another

"If a two-dimensional creature ate something it could not digest completely, it would have to bring up the remains the same way it swallowed them, because if there were a passage right through its body, it would divide the creature into two separate halves..."

Stephen W. Hawking, in *A Brief History of Time*

AMONG NATURE'S FASCINATING MOLECULAR MACHINES, few are as elegant, or as ingenious, as a type II DNA topoisomerase. In a reaction catalyzed by this enzyme, a DNA double helix slices through the entire interface between two halves of the dimeric enzyme, as well as through a second DNA double helix straddling the two halves. Yet the passage of this DNA "knife" leaves no trace of a chemical record within the enzyme–DNA complex. Would not the enzyme–DNA complex, like a two-dimensional creature with a digestive tract bisecting its body, be in danger of falling apart whenever a DNA double helix slices through it? As we shall see in Chapter 10, there may be some danger of this, but under normal conditions the risk is quite minimal. Stephen Hawking, a great mind and renowned physicist and cosmologist, had neglected to point out the obvious: A two-dimensional creature with a digestive channel running through it can stay forever in one piece if it refrains from opening both ports at the same time!

A type II DNA topoisomerase, by definition, cleaves and rejoins both strands of a DNA double helix during its actions on DNA, and is designated by an even Roman numeral for historical reasons; very closely related enzymes are further distinguished by Greek letters. Some of the better-studied examples are yeast DNA topoisomerase II, human DNA topoisomerase IIα and IIβ, and *Escherichia coli* DNA topoisomerase IV.

In more recent years, these members have also been joined by DNA topoisomerase VI of *Sulfolobus shibatae*, an archaeal microorganism that lives in hot springs,[1] and by DNA topoisomerase VI of *Arabidopsis thaliana*, which was an obscure weed before it came to serve as a model organism in molecular genetics studies in the late 20th century. Some type II enzymes may also go by different names. Examples are phage T4 DNA topoisomerase, which lacks a numerical designation because the bacterial virus encodes no other DNA topoisomerases, and *E. coli* DNA gyrase (or simply *E. coli* gyrase), which has the formal name *E. coli* DNA topoisomerase II, but has retained the original name bequeathed at the time of its discovery.

The gyrase of *E. coli*, discovered in 1976,[2] is a remarkable enzyme that catalyzes the ATP-dependent negative supercoiling of a relaxed DNA ring. Supercoiling of a relaxed DNA contorts and strains it, and the reaction is energetically unfavorable (recall the discussion in Chapter 2 of energetically favorable and unfavorable reactions). The *E. coli* gyrase manages to overcome this unfavorable energy barrier, however, by coupling its supercoiling reaction to the energetically favorable hydrolysis of ATP to ADP and orthophosphate (orthophosphate is often termed "inorganic phosphate" by biochemists, and given the notation P_i). Thus gyrase, as well as all other members of the type II DNA topoisomerase family, is actually two enzymes in one: a DNA topoisomerase and a DNA-dependent ATPase that converts ATP to ADP and P_i.

The discovery of gyrase provides yet another example of serendipity in science. In the 1970s, Kiyoshi Mizuuchi and Howard Nash were studying site-specific recombination catalyzed by the enzyme phage λ integrase. As we saw in Chapter 4, the phage enzyme, together with a protein encoded by *E. coli*, cleaves a total of four DNA strands within two unique sequences denoted by POP′ and BOB′, and rejoins all of the broken strands to give POB′ and BOP′ (see Fig. 4-6). Mizuuchi and Nash created a single double-stranded DNA molecule carrying both the POP′ and BOB′ sequences. When they used this DNA as a substrate in the presence of purified λ integrase and an extract of *E. coli* cells, they noticed that the reaction had a curious ATP requirement: ATP was needed if the input DNA was in the form of a nicked ring, but not if the DNA ring was in the negatively supercoiled form.[3] This finding hinted that the integrase reaction was somehow enhanced in a negatively supercoiled DNA, and that ATP was somehow required to convert DNA into the negatively supercoiled configuration. Indeed, Mizuuchi and Nash also observed that a nicked DNA ring was not only converted to a covalently closed form, presumably owing to the presence of DNA ligase in the cell extract, but that when ATP was added to the assay mixture, the DNA ring also became negatively supercoiled. It therefore appeared that negative super-

1. As indicated in earlier chapters, all living organisms can be classified into three domains: bacteria, eukarya, and archaea. See Woese, C.R. and Fox, G.E. 1977. *Proc. Natl. Acad. Sci.* **74:** 5088–5090.
2. Gellert, M., et al. 1976. *Proc. Natl. Acad. Sci.* **73:** 3872–3876.
3. Mizuuchi, K. and Nash, H. 1976. *Proc. Natl. Acad. Sci.* **73:** 3524–3528.

coiling of the DNA was caused either by the binding of certain protein molecules in the cell extract in an ATP-dependent way, or by the catalytic action of an ATP-dependent enzyme in the cell extract. But coiling of a DNA by the ATP-dependent binding of protein molecules that change the twist and writhe of a DNA in space could not readily explain another striking observation made by Mizuuchi and Nash: If the input DNA was in a linear form, no integrase-promoted recombination was observed, whether ATP was present or absent. Taken together, these observations pointed to the presence of an enzymatic activity in the cell extract that catalyzes the ATP-dependent negative supercoiling of a DNA ring. In pursuit of evidence, Mizuuchi and Nash joined forces with Martin Gellert to purify this postulated supercoiling activity, and so established gyrase as an ATP-dependent DNA supercoiling enzyme.[2]

For several reasons, the discovery of gyrase greatly increased interest in DNA topology and DNA topoisomerases. First, the reaction catalyzed by gyrase is itself quite fascinating—the contortion of a long DNA by a miniature rope-making machine (Fig. 5-1). Second, the finding immediately suggested a plausible answer to a question of long standing that was considered in Chapter 2: Why are DNA rings from natural sources negatively supercoiled?[4] Before the discovery of gyrase, very few researchers had considered the possibility that an enzyme might actively coil up a DNA mole-

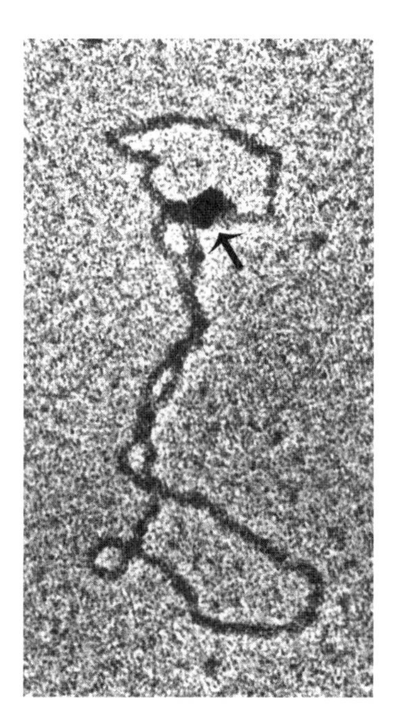

Figure 5-1. Electron micrograph showing a gyrase (marked by an arrow) bound to a supercoiled DNA. (Reprinted, with permission, from Wang, J.C. 1987. *Harvey Lect.* **81:** 91–110.)

4. As we saw in Chapter 2, this very question had a close tie to the discovery of *E. coli* DNA topoisomerase I.

cule.[5] Third, this discovery lent credence to the biological importance of DNA supercoiling, a subject to be discussed in Chapter 9, with the strong dependence of the λ integrase-catalyzed reaction on DNA negative supercoiling as a striking example. Fourth, shortly after the discovery of gyrase, the genes *gyrA* and *gyrB* (encoding its two subunits) were found to be identical to the genes *nalA* and *cou*, respectively. Mutations within *nalA* and *cou* were known to be responsible for resistance to antibiotics of the nalidixate and coumermycin families, respectively. These two antibiotics were also known to strongly inhibit DNA replication, and the identification of *nalA* and *cou* as *gyrA* and *gyrB*, respectively, not only demonstrated the importance of gyrase in DNA replication, but also showed that this enzyme is the target of the two families of antibiotics. At that time, two and only two drugs were actually tested for their effects on gyrase, and both were found to target this enzyme. If only all drug discovery programs were as effective! Many more drugs targeting gyrase and other DNA topoisomerases are now known, a topic to be discussed in Chapter 10.

A NEW SUBFAMILY OF DNA TOPOISOMERASES: TYPE II

In the absence of ATP, gyrase can slowly relax negatively supercoiled DNA. Thus, early on, gyrase was assumed to act just like other DNA topoisomerases then known, all of which were thought to transiently break a DNA double helix one strand of at a time (see Chapters 3 and 4). The ability of gyrase to supercoil a DNA ring was assumed to be closely tied to its ATPase activity, but the earlier thoughts about its mechanism of action all involved reversible breakage of only one DNA strand. In 1979, however, Patrick O. Brown and Nicholas R. Cozzarelli (1938–2006) showed that gyrase actually acted by a very different mechanism: It transiently breaks both strands of a DNA double helix and moves another DNA double helix through this double-stranded break in the DNA.[6] Around the same time, the same mechanism was independently proposed by Leroy F. Liu, Chung-Cheng Liu, and Bruce M. Alberts for phage T4 DNA topoisomerase.[7,8] The simple yet elegant experiments of Brown and Cozzarelli and of

5. This possibility was first raised by Huan Lee, a physicist friend of mine. When I first told him about the *E. coli* ω protein in 1970, Lee remarked that, in analogy to the existence of opposing pairs of particles like the electron and positron and the proton and anti-proton, the existence of an ω protein that relaxes negatively supercoiled DNA might be suggestive of the existence of an anti-ω protein that negatively supercoils a relaxed DNA; he further argued that the higher energy state of a supercoiled DNA would impose an ATP requirement for such an anti-ω reaction. (See Wang, J.C. 2004. In *DNA topoisomerases in cancer therapy: Present and future* [ed. T. Andoh], pp. 1–13. Kluwer/Plenum, New York.)

6. Brown, P.O. and Cozzarelli, N.R. 1979. *Science* **206:** 1081–1083.

7. Liu, L.F., et al. 1979. *Nature* **281:** 456–461. This paper was the first to introduce the idea of a DNA topoisomerase that catalyzes the transport of one DNA double helix through a transient double-stranded break in another.

8. Liu, L.F., et al. 1980. *Cell* **19:** 697–707.

Liu et al. that led to this conclusion required the use of a DNA ring with a unique linking number, *Lk* (see Chapter 2 for discussions of *Lk*).[6,8] The preparation of this substrate became feasible through the work of Walter Keller in 1975. When he analyzed an SV40 DNA preparation by electrophoresis through an agarose gel, Keller made a surprising observation: The preparation formed a ladder of sharp DNA bands (Fig. 5-2, middle and right lanes).[9] The formation of this DNA ladder was traced to a chance encounter, in an extract of SV40-infected cells, between supercoiled SV40 DNA and a DNA topoisomerase (a type IB enzyme, as described in Chapter 4). As a result of this encounter, the negatively supercoiled SV40 DNA rings become partly relaxed. Other than their *linking numbers*, the partly relaxed DNA rings are identical, and they are thus topoisomers (as we saw in Chapter 2). Owing to their differences in spatial coiling, and therefore their average three-dimensional shape, many of these topoisomers migrate at different speeds through the gel matrix to form a ladder of bands. But if two topoisomers are both highly supercoiled, then each resembles the shape of a twisted rod, and gel electrophoresis fails to resolve such a rod-shaped pair. We would expect

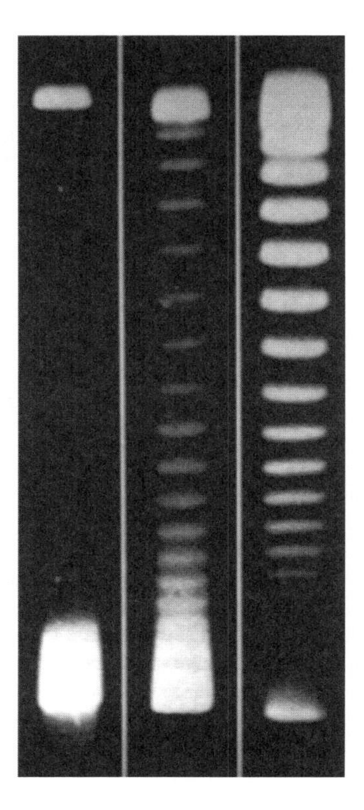

Figure 5-2. Separation of SV40 DNA rings of different linking numbers by electrophoresis through an agarose gel. In the *left* sample, untreated with topoisomerase, the broad band near the bottom of the photograph contains negatively supercoiled rings, which migrate much faster than the small amount of nicked DNA rings (band near top of the sample). The *middle* and *right* samples were treated with the same amount of the enzyme for 5 and 30 min, respectively. Partial relaxation of the supercoiled DNA yields a ladder of sharp bands migrating between the highly negatively supercoiled DNA and the nicked DNA. (Reprinted, with permission, from Keller, W. 1975. *Proc. Natl. Acad. Sci.* **72:** 4876–4880.)

9. Keller, W. 1975. *Proc. Natl. Acad. Sci.* **72:** 4876–4880.

therefore that all highly supercoiled topoisomers would bunch up and migrate as a thick band ahead of the sharp individual bands of the resolved topoisomers of discrete linking numbers (Fig. 5-2). Subsequent experiments have shown that, in the region where different topoisomers are well resolved in the gel, the DNA topoisomeres in two adjacent bands in the ladder differ by only 1 in their linking numbers. This discrete resolution of topoisomers with different linking numbers is quite remarkable; the average value of *Lk* of an SV40 DNA ring is ~450, yet a 1/450 or 0.2% difference between the linking numbers of two topoisomers is sufficient for their physical separation.

Keller's accidental discovery introduced a simple yet powerful method for the study of the topology of DNA rings. One of the earlier experiments using this approach was carried out in Jerome Vinograd's laboratory shortly after Keller's work.[10] A DNA band with a unique *Lk* was first isolated from a topoisomer ladder by recovering the DNA from the gel slice. The isolated DNA was then treated with a type IB DNA topoisomerase, and the product was analyzed again by gel electrophoresis. Treatment with the topoisomerase split the single DNA band into a group of discrete bands, with a difference in *Lk* of 1 between adjacent bands.[10] This amazing further splitting of a single band was actually anticipated: The presence of a DNA topoisomerase would alleviate the invariance in *Lk*, and thus allow formation of a population of DNA topoisomers that do not differ greatly in their energies. Indeed, it had been shown that when a nicked DNA ring was treated with DNA ligase to seal the nick, a topoisomer ladder, called a Boltzmann population, was formed (Fig. 5-3).[11]

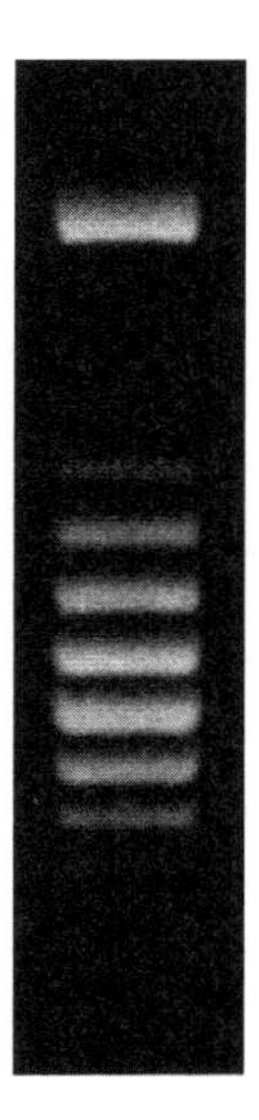

Figure 5-3. Treatment of a nicked DNA ring yields a group of topoisomers that migrate differently through an agarose gel. A DNA ring of ~10,000 bp was first treated with pancreatic DNase I to introduce a few nicks in each ring, and the nicked DNA was then treated with DNA ligase. The nicked DNA band is seen near the top of the gel, and the group of bands migrating below it represent a Boltzmann distribution of topoisomers with different linking numbers formed when the nicks were sealed.

10. Pulleyblank, D.E., et al. 1975. *Proc. Natl. Acad. Sci.* **72:**4280–4284.

11. Depew, R.E. and Wang, J.C. 1975. *Proc. Natl. Acad. Sci.* **72:** 4275–4279.

A population of molecules in thermal equilibrium is called a Boltzmann population after the Austrian physicist Ludwig Boltzmann (1844–1906). Treatment of a DNA ring with a type IB DNA topoisomerase, or treatment of a nicked DNA ring with DNA ligase, yields a Boltzmann population of DNA topoisomers that differ only in their linking numbers. Boltzmann populations of molecules are ever present in nature, but it is rare that individual members of such a population can be physically separated and visualized, because they readily interconvert at ordinary temperatures. In the case of the DNA topoisomers that differ in their linking numbers, however, interconversion from one to another is possible only when the DNA rings are temporarily nicked. Thus, resealing of a nicked DNA ring by DNA ligase, or the removal of a type IB DNA topoisomerase from a solution of DNA rings, has the net effect of permanently trapping a Boltzmann population of DNA linking number topoisomers: they can no longer interconvert because of the topological invariance of the linking number.

To understand why, as seen in Figure 5-3, all covalently closed DNA rings migrate significantly faster than the nicked ring, it is helpful to think in terms of Lk, Lk^0, and their difference ($Lk - Lk^0$). Before DNA ligase treatment, the DNA rings are unconstrained, and upon sealing of the nicks, the average Lk of the covalently closed product is expected to be equal to Lk^0_{lig}, where the subscript specifies that it is the linking number of the DNA ring in its most stable structure under the DNA ligase treatment conditions. Lk remains unchanged during subsequent gel electrophoresis, but the value of Lk^0 of the DNA ring is altered slightly to become Lk^0_{elec}, because of a very small change in the helical structure of DNA between the conditions of DNA ligase treatment and the conditions of electrophoresis. In fact, Lk^0_{elec} is actually a bit lower than Lk^0_{lig}, and thus the average linking difference ΔLk of the covalently closed DNA rings during electrophoresis ($Lk - Lk^0_{elec}$), which is equal to ($Lk^0_{lig} - Lk^0_{elec}$), is greater than 0. Consequently, the covalently closed topoisomers are slightly positively supercoiled under the conditions of electrophoresis and migrate significantly faster than the nicked DNA ring.

Remarkably, when a DNA ring with a unique Lk was treated with gyrase or phage T4 DNA topoisomerase, Lk was found to change by even but not by odd numbers.[6,8] These findings led Brown and Cozzarelli and Liu et al. to conclude that gyrase and phage T4 DNA topoisomerase catalyze topological transformations of DNA rings by transiently breaking both strands of a DNA double helix to pass another DNA double helix through the break. A simple ribbon-cutting exercise serves to demonstrate why the passage of one DNA segment through another on the same double-stranded DNA ring leads to an even change in Lk. When the two ends of a ribbon are joined to form a simple ring, the two edges of the ring are not linked ($Lk = 0$), and the circular ribbon would split into two separate rings if it were cut longitudinally. But if a cut is first made across the circular ribbon and a different part of the same ribbon is passed through this

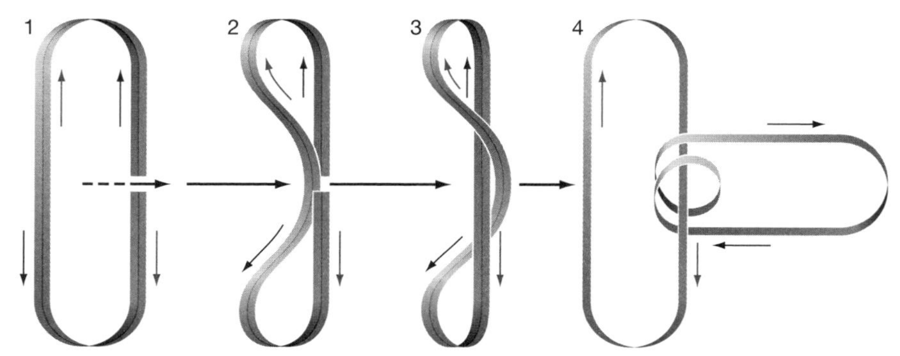

Figure 5-4. A ribbon-cutting exercise demonstrating a change of 2 in the *Lk* value of a DNA ring when one segment of the DNA passes a transient break in another. In the *leftmost* image (*1*), a circular ribbon is cut at one place for the passage of another part of the same ribbon through it. The two edges of the ribbon (shown in black and blue) represent two strands, with their polarities marked by the arrows along the edges. The two edges (strands) of the ribbon in *1* are not linked in the original ribbon. Passage of another part of the ribbon through the cut (*2*), followed by resealing of the cut, leads to the formation of the twisted ribbon shown in *3*. The two edges are now doubly linked ($Lk = 2$), as can be seen by slitting the ribbon longitudinally (*4*). By following the conventions described in the legend to Fig. 2-2, we see that in the case shown, *Lk* is changed from 0 to +2 (which can also be seen from the left-handed plectonemic coiling of the ribbon drawn in *3*). (Redrawn, with permission, from Wang, J.C. 1982. *Sci. Am.* **247:** 94–109.)

cut before the severed ribbon is rejoined, then the two edges of the circular ribbon will become doubly intertwined: if the ribbon is again split longitudinally into two, one member of the pair can be seen to revolve twice around the other (Fig. 5-4).

In contrast to the ATP-dependent DNA negative supercoiling reaction catalyzed by *E. coli* gyrase, the phage T4 DNA topoisomerase catalyzes not the supercoiling of a relaxed DNA ring, but rather ATP-dependent relaxation of either positively or negatively supercoiled DNA rings.[7,8] This ability of T4 DNA topoisomerase to relax supercoiled DNAs, but not to promote supercoiling, was soon found to be shared by similar activities in eukaryotic organisms ranging from yeast to human.[12] These eukaryotic enzymes, all named DNA topoisomerase II, are proteins with two identical copies of a single polypeptide. By contrast, each gyrase is a dimeric molecule represented by $(BA)_2$, in which B and A denote, respectively, the GyrB and GyrA polypeptide, and each T4 enzyme contains three different polypeptides, two copies of each encoded by the phage genes *39*, *52*, and *60*. Enzymologists long ago coined the special term "protomers" for identical protein molecules in an enzyme. Thus, gyrase is made of two GyrA and two GyrB protomers, and T4 DNA topoisomerase of two protomers of each of the gene *39*,

12. Baldi, M.I., et al. 1980. *Cell* **20:** 461–467; Hsieh, T. and Brutlag, D. 1980. *Cell* **21:** 115–125; Miller, K.G., et al. 1981. *J. Biol. Chem.* **256:** 9334–9339; Goto, T. and Wang, J.C. 1982. *J. Biol. Chem.* **257:** 5866–5872.

52, and *60* protein products. When the amino acid sequence of the single-polypeptide yeast DNA topoisomerase II (Top2) was deduced from the nucleotide sequence of its gene (*TOP2*), the first or amino-terminal half of the sequence was found to be similar to the sequence of *E. coli* GyrB, and the second half similar to that of GyrA. The sequences of the three T4 DNA topoisomerase subunits were also shown to have corresponding stretches in the single yeast Top2 polypeptide. Clearly, all of these enzymes are closely related, despite differences in their subunit structures or their abilities to supercoil or relax DNA rings: they constitute a new subfamily of Type II DNA topoisomerases. These Type II enzymes were further distinguished as type IIA, after the 1997 discovery of the *S. shibatae* DNA topoisomerase VI, a type IIB enzyme that differs significantly from the type IIA enzymes.[13] Among all type II enzymes, however, bacterial gyrase remains the black sheep that catalyzes negative supercoiling, whereas all the others can only relax supercoiled DNA in the presence of ATP. Why does gyrase differ from the other members of the type II family of DNA topoisomerases? We shall address this question in a later section of the chapter.

LIFTING A MENTAL BLOCK

Three years elapsed between the discovery of *E. coli* gyrase and the realization that it catalyzes the transient breakage of both strands of a DNA double helix. With hindsight, there were already strong hints of this mechanism of action of the gyrase shortly after its discovery. In 1977, several investigators observed that, in the presence of thr gyrase-targeting drug nalidixate (also referred to as nalidixic acid), the addition of a detergent, sodium dodecyl sulfate (SDS), would lead to the gyrase-mediated conversion of a DNA ring to a linear form.[14] Further, it appeared that this conversion was accompanied by the formation of a very stable (probably covalent) protein–DNA complex at the end of each broken DNA strand.[14] It now seems to us surprising that these results did not immediately suggest that gyrase could catalyze the transient cleavage of both DNA strands of a double helix. Why was this logical conclusion missed?

The severance of a DNA double helix had long been thought to be a dreadful cellular event. It was also well known that intricate cellular machineries had evolved to repair such double-stranded breaks (DSBs), and that without such repair functions, the cells would perish. Thus there was perhaps a mental block against the idea that a normal cellular activity like DNA gyrase could make a living by constantly generating DSBs, even transient ones. Would not the broken DNA ends, like the partitioned body parts in a two dimensional animal, fly apart? Studies of gyrase and T4 DNA topoisomerase in 1979 and 1980 finally lifted this mental block. Eons ago, as the DNA world was

13. Bergerat, A., et al. 1997. *Nature* **386:** 414–417.

14. Sugino, A., et al. 1977. *Proc. Natl. Acad. Sci.* **74:** 4767–4771; Gellert, M., et al. 1977. *Proc. Natl. Acad. Sci.* **74:** 4772–4776.

prospering, the type II DNA topoisomerases had apparently emerged to meet the challenge of the DNA entanglement problem (posed in Chapter 1), and these enzymes probably evolved to minimize the inherent risk of broken DNA ends flying apart. As we shall see below and in Chapters 6–8, the type II DNA topoisomerases manage to do so splendidly and with style, though not with complete impunity (Chapters 7 and 10).

A TYPE II DNA TOPOISOMERASE AS AN ATP-DEPENDENT PROTEIN CLAMP

Soon after the discovery of gyrase and other type IIA DNA topoisomerases, it became clear that one single dimeric enzyme molecule was sufficient to perform the remarkable feat of repeatedly moving one DNA double helix through another. How is this possible?

There are four interrelated parts to this seemingly simple question. First, the enzyme must break the double helix of the enzyme-bound DNA, and, before the end of a reaction cycle, reseal it. Second, after unlocking it, the enzyme must open the "gate" in a bound DNA segment. That is, after transiently breaking the DNA double helix, the enzyme must pull apart the DNA ends now attached to the two halves of the enzyme, to form a gate in the DNA wide enough for a second DNA double helix to pass through. The pair of DNA double helices involved in this reaction may reside on the same DNA molecule or on two separate molecules. For ease of discussion, the enzyme-bound DNA double helix containing the DNA gate has been termed the gate- or G-segment, and the DNA double helix to be transported through the G-segment named the transported- or T-segment. Third, after passage of the T-segment through the G-segment, the T-segment DNA must exit the enzyme without passing back through the G-segment, because entering and then backing out of the same DNA gate would accomplish nothing. Fourth, one or more of the steps described so far must be coupled to ATP usage. We shall consider these four aspects in more detail here and in the next two sections.

The first of the four issues raised above is probably the simplest, because earlier studies of the type I DNA topoisomerases had already shown that DNA breakage and rejoining could be accomplished by transesterification reactions between an enzyme and DNA (as discussed in Chapters 2–4). For a type II enzyme, the participation of a pair of active-site tyrosyl residues, one in each half of the dimeric enzyme, would seem sufficient to unlock a DNA gate. Indeed, Tyr-122 of the GyrA-subunit of *E. coli* gyrase, and Tyr-782 of yeast DNA topoisomerase II, have been shown by direct biochemical experiments to be the active-site tyrosines.[15] Moreover, the corresponding tyrosines in all other known type IIA DNA topoisomerases have subsequently been deduced by comparison of their amino acid sequences with those of the *E. coli* and yeast enzymes.[15] When we replaced the active-site tyrosine of a

15. Horowitz, D.S. and Wang, J.C. 1987. *J. Biol. Chem.* **262:** 5339–5344; Worland, S.T. and Wang, J.C. 1989. *J. Biol. Chem.* **264:** 4412–4416.

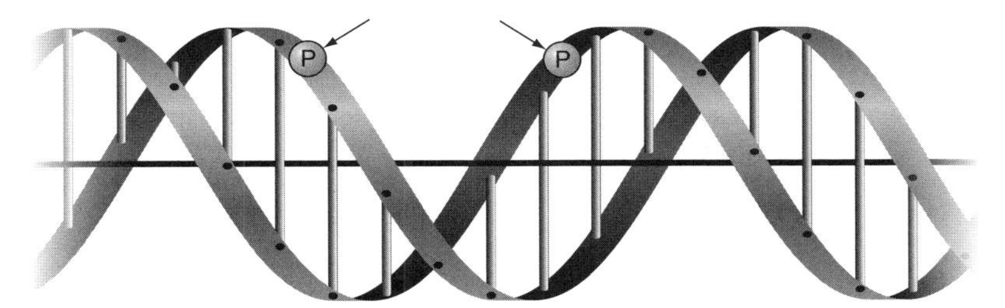

Figure 5-5. A model of the DNA double helix showing the approximate positions of two phosphorus atoms (P) in a pair of DNA strands. The phosphorus atoms, where transesterification between the DNA and a type IIA DNA topoisomerase occurs, are staggered by four base pairs. Cleavage at these positions, inside a pair of catalytic pockets of a type IIA DNA topoisomerase, gives a pair of broken DNA ends with 5′ single-stranded overhangs, each four nucleotides long. The 5′-phosphoryl groups at the tips of the single-stranded overhangs are covalently linked to a pair of active-site tyrosyl residues, one in each half of a dimeric type IIA DNA topoisomerase.

yeast enzyme with a phenylalanine, the resulting mutant enzyme was invariably incapable of cleaving a DNA ring or changing its linking number. As mentioned in the preceding chapter, phenylalanine is very similar to tyrosine but lacks the catalytically important hydroxyl group of tyrosine. Furthermore, we showed that a yeast DNA topoisomerase II "heterodimer" with one normal (wild-type) polypeptide and one Y782F mutant polypeptide (Y782F denotes the replacement of Tyr-782 by a phenylalanine, Y and F being the respective one-letter amino acid codes for tyrosine and phenylalanine; similar notations will be used for other mutations) exhibits no DNA relaxation activity, but is still capable of cleaving one DNA strand and forming a covalent link between the wild-type polypeptide and the 5′ phosphoryl end of the broken DNA strand.[16] All of these experiments indicate that the type IIA and type IA enzymes use the same transesterification chemistry for the breakage and rejoining of DNA strands. In other words, in a type IIA enzyme a pair of symmetry-related tyrosyl residues, one in each half of the dimeric enzyme, is directly involved in the transient breakage of a DNA double helix. Mapping of the DNA cleavage sites of various type IIA DNA topoisomerases has shown that each pair of transient breaks in a DNA double helix is separated by four base pairs. Presumably, the pair of active-site tyrosyl residues is related by a molecular dyad that also bisects the two target phosphorus atoms separated by four base pairs (see Fig. 5-5).[17]

The second question is how, after unlocking the DNA gate by cleaving the DNA strands of a double helix, does a type II enzyme open the gate for a T-segment to move through it? The diameter of a DNA double helix is about 2 nm (20 Å); there-

16. Liu, Q. and Wang, J.C. 1998. *J. Biol. Chem.* **273:** 20252–20260.

17. Reviewed in Wang, J.C. 1996. *Annu. Rev. Biochem.* **65:** 635–692.

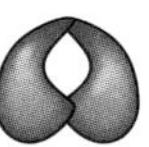

Figure 5-6. A model in which the two halves of the dimeric type IIA DNA topoisomerase enzyme form an ATP-modulated protein clamp. The protein clamp closes when ATP binding triggers a conformational change in it, and reopens upon hydrolysis of the bound ATP and release of ADP and P_i. An analog of ATP that is not hydrolyzed by the enzyme, ADPNP, can also bind to the enzyme ATPase pockets and trigger closure of the protein clamp, but in this case the clamp can no longer reopen.

fore, the DNA gate must be 2 nm or wider when fully opened. Clearly then, the enzyme–DNA complex must undergo large conformational changes in order to open and close the DNA gate. How might these conformational changes occur?

A clue came to us in 1992.[18] Because all members of the type IIA subfamily appear to march to the same drumbeat in terms of basic reaction steps, their mechanistic features are most likely shared by all members. In 1992 we found that, in the presence or absence of ATP, the yeast type IIA enzyme can bind DNA of any shape: linear, nicked, relaxed, or supercoiled. However, preincubation of the enzyme with the nonhydrolyzable ATP analog ADPNP triggers a remarkable change: the ADPNP-bound enzyme can no longer bind any form of ring-shaped DNA, but can still bind linear DNA. Furthermore, we observed that a bound linear DNA readily dissociates from the enzyme–ADPNP complex if the salt concentration in the solution is increased to reduce the affinity of the enzyme for DNA. This dissociation is prevented, however, if the ends of the bound linear DNA are first joined by treatment with a DNA ligase. (Note that in our original experiments we used a bacterial DNA ligase that requires NAD rather than ATP as a cofactor, to avoid the potential complication that the presence of ATP, might displace the enzyme-bound ADPNP and thus obscure the experimental outcome).[18,19]

These observations led us to the idea that a dimeric yeast Top2 acts like a molecular clamp with a large central hole: binding of ATP to the enzyme would close the clamp to give a doughnut-shaped molecule, and hydrolysis of the bound ATP and release of the hydrolytic products would reopen the clamp. Incubation of the enzyme with a nonhydrolyzable ATP analog, on the other hand, would permanently close the enzyme clamp (Fig. 5-6). Presumably then, when the protein clamp closes, a large central hole is formed; the size of this hole is sufficient for a linear DNA molecule to thread through, but not quite large enough for the entrance of a ring-shaped DNA. Hence, preincubation of the Top2 enzyme with ADPNP would convert it into a form that binds only linear DNA. If, however, the ends of a linear DNA were joined after it had threaded through a protein ring,

18. Roca, J. and Wang, J.C. 1992. *Cell* **71:** 833–840.

19. It was observed several years earlier by Neil Osheroff that in the presence of ADPNP, *Drosophila* DNA topoisomerase II would remain bound to a DNA ring even at very high salt concentrations, but not to the same DNA in a linear form (Osheroff, N. 1986. *J. Biol. Chem.* **261:** 9944–9950).

Figure 5-7. A model in which a type IIA DNA topoisomerase catalyzes the transport of one DNA double helix through another. The protein clamp bound to a DNA segment admits a second DNA segment while in the open state; closure of the clamp upon ATP binding traps the second DNA segment and moves it through the first. The rods labeled G and T represent the first and second DNA segments, respectively; a transient double-stranded break within the G- or gate-segment allows passage of the T- or transport-segment through the G-segment. (Redrawn, with permission of Elsevier, from Roca, J. and Wang, J.C. 1992. *Cell* **71:** 833–840.)

then the DNA ring formed would encircle the doughnut-shaped protein, and the two topologically linked molecules could no longer come apart, even in the presence of excess salt. Therefore, a type IIA enzyme can be modeled as an ATP-operated protein clamp (Fig. 5-6). This model immediately suggested that ATP could be utilized to effect large movements between the two halves of a dimeric DNA topoisomerase II, and that such movements could in turn be coupled to the opening and closing of the DNA gate.

A year before these DNA-binding experiments were conducted, biochemical and X-ray diffraction studies had shown that the pair of ATPase domains of *E. coli* gyrase, each located in the amino-terminal half of a GyrB protomer, would form a dimer in the presence of ADPNP.[20] Therefore, the pair of ATPase domains of this enzyme likely constitute an entrance port, which closes when ATP binds these domains, and reopens upon hydrolysis of bound ATP and/or release of the hydrolytic products.[21] This putative entrance gate has been termed the N-gate, because in yeast and other eukaryotic DNA topoisomerase II enzymes the ATPase domain is near the amino- (or N-)terminal region of each polypeptide. Presumably, the N-gate must open for the binding of the enzyme to a DNA ring, or for the entrance of a T-segment into the interior of a DNA-bound enzyme. The clamp model further assumes that if ATP binds to a DNA-bound enzyme after the entrance of the T-segment, then closure of the N-gate would trap the T-segment and facilitate its transport through the G-segment.[18] Such a scheme is illustrated in Figure 5-7.

20. Wigley, D.B., et al. 1991. *Nature* **351:** 624–629. The ATPase domain of yeast DNA topoisomerase II was determined a few years later (Classen, S., et al. 2003. *Proc. Natl. Acad. Sci.* **100:** 10629–10634).

21. Many biological macromolecules use the binding of ATP, guanosine triphosphate (GTP), or other small molecules to effect large conformational changes. In the case of the ATPase domain of a type II DNA topoisomerase, or similar domains in proteins like MutL (involved in DNA repair), the polypeptide is not properly folded for dimer formation in the absence of ATP. Binding of ATP to the polypeptide, however, alters its folding and forms a good surface for dimerization (see Corbett, K.D. and Berger, J.M. 2003. *EMBO J.* **22:** 151–163; Ban, C. and Yang, W. 1998. *Cell* **95:** 541–552.). Nature has also learned how to avoid wasting a bound ATP by prematurely hydrolyzing it: Catalysis of ATP hydrolysis requires the cooperation of amino acid side chains from *both* halves of a type II DNA topoisomerase. Thus a bound ATP can be hydrolyzed only after it has accomplished its mission of inducing dimerization (see Footnotes 16 and 20).

AN ENZYME OF TWO GATES

How might a captured T-segment move through the G-segment of an enzyme-bound DNA double helix and exit the protein clamp? According to the clamp model, when the N-gate of a DNA-bound type II enzyme closes to trap a T-segment, the entrance port holding the T-segment is in a strained configuration. This strain in turn triggers a conformational change in the enzyme–DNA complex, leading to the opening of the DNA gate and the passage of the T-segment into the large central cavity in the enzyme. Once the T-segment has sailed through the G-segment and become moored in a temporary docking area, there are basically two ways in which the transported T-segment can emerge from the enzyme interior (shown in Fig. 5-8). In one model, the T-segment could exit the enzyme again through the N-gate, after its reopening following ATP hydrolysis and release of the hydrolytic products. But in this case, at least one side of the G-segment must first detach from the enzyme to permit backtracking of the T-segment without its again passing through the G-segment (Fig. 5-8a). Such a model requires only one protein gate and is therefore termed the one-gate model. The second model postulates that, during a complete reaction cycle, the T-segment can pass unidirectionally

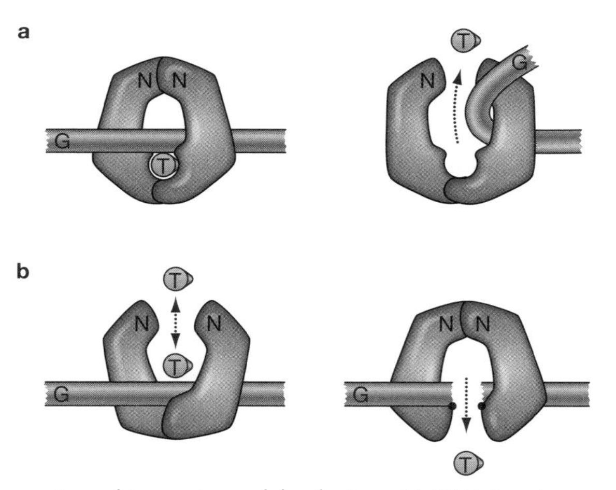

Figure 5-8. The one-gate and two-gate models of a type IIA DNA topoisomerase. The jaws of the ATPase domains in the protein clamp are labeled N to signify their likely presence at the amino-terminal parts of the dimeric enzyme. The T-segment, labeled T, is represented by an end view of it as a circle, and the G-segment is labeled G. (*a*) In the one-gate model, the T-segment enters the G-segment bound enzyme and moves through the DNA gate. When the enzyme clamp reopens, the G-segment, or a part of it, dissociates from the enzyme to allow the exit of the moored T-segment. (*b*) In the two-gate model, the T-segment can go in and out of a G-segment-bound protein clamp when the clamp is in its open state (*left* drawing). As illustrated on the *right*, a second protein gate on the opposite side of the entrance gate opens to allow the exit of the T-segment. (Redrawn, with permission, from Roca, J. and Wang, J.C. 1994. *Cell* **77:** 609–616.)

through a type IIA DNA topoisomerase, first entering the N-gate, then sailing through the transiently opened DNA gate in the G-segment, and finally exiting a second protein gate opposite the N-gate (Fig. 5-8b). This—the two-gate model—requires that there be a separate DNA entrance and exit gate in each dimeric enzyme molecule.

The two-gate model was proposed independently by two groups of researchers shortly after the finding that a type II DNA topoisomerase acts by transporting one DNA double helix through another.[22] The model was not well received initially, probably for the same reason discussed earlier: fear of the two enzyme halves coming apart while they hold the two broken DNA ends. In the case of *E. coli* gyrase, there was actually a very good reason for proposing a two-gate rather than a one-gate model: We knew that gyrase interacts with a large stretch within a G-segment that spans some 140 bp.[23] The one-gate model would require at least half of this stretch to detach from the enzyme, involving extensive disruptions of protein–DNA interactions. Perhaps it would be more energetically economical to disrupt a protein–protein interface that constitutes the postulated exit gate, rather than disrupt a large protein–DNA interface.

In 1994, 14 years after the two-gate model had initially been suggested, we devised an experimental test for it.[24] The experiment was fairly simple conceptually. We used as the starting DNA a pair of singly interlocked DNA rings that could become unlinked upon a single passage of one ring through the other. A yeast DNA topoisomerase II (Top2) molecule was bound to one of the component DNA rings, and this binding defined the G-segment. The nonhydrolyzable ATP analogue ADPNP was then added to trigger permanent closure of the N-gate of the bound enzyme, as well as passage of the T-segment, trapped inside the enzyme clamp, through the G-segment. If the T-segment resided on a separate DNA ring (not bearing the G-segment), then the pair of interlocked rings would come apart when the G-segment crossed the T-segment. As seen in Figure 5-9, the one-gate and two-gate models predict very different outcomes of this decatenation, in terms of which DNA rings would be retained within the closed protein clamp. According to the one-gate model, because the N-gate cannot reopen in the presence of the bound ADPNP, both unlinked rings would be expected to remain locked within the closed protein clamp. In the two-gate model, on the other hand, the unlinked ring containing the T-segment would simply pass through the exit gate into the solution, leaving only one ring (containing the G-segment) inside the locked protein clamp. The results of this experiment were very clear: The enzyme acts according to the two-gate model![24]

22. Mizuuchi, K., et al. 1980. *Proc. Natl. Acad. Sci.* **77:** 1847–1851; Wang, J.C., et al. 1981. In *Mechanistic studies of DNA replication and genetic recombination* (ed. B.M. Alberts and C.F. Fox), pp. 769–784. Academic Press, New York.

23. Liu, L.F. and Wang, J.C. 1978. *Proc. Natl. Acad. Sci.* **75:** 2098–2102; **1978.** *Cell* **15:** 979–984. A slightly shorter protected region of 128 bp with a 13-bp central segment that was most strongly protected was implicated when DNA backbone bonds in a gyrase-bound DNA were probed with the highly reactive hydroxyl free radical (Orphanides, G. and Maxwell, A. 1994. *Nucleic Acids Res.* **22:** 1567–1575).

24. Roca, J. and Wang, J.C. 1994. *Cell* **77:** 609–616.

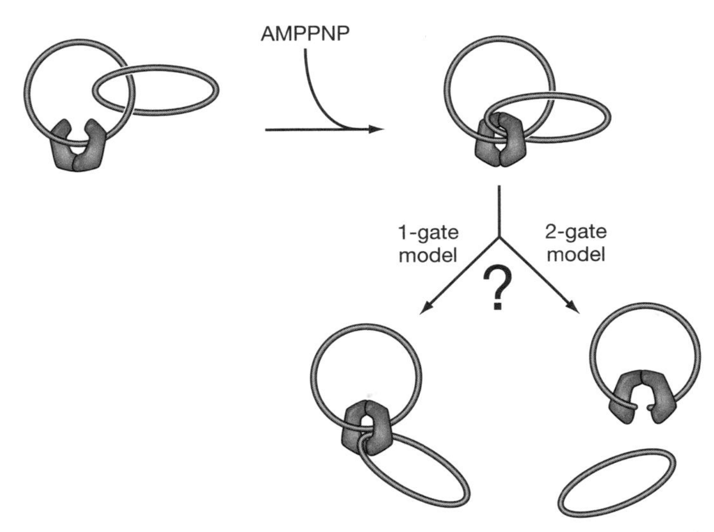

Figure 5-9. Design of an experiment to test the one- and two-gate models illustrated in Fig. 5-8. A type IIA DNA topoisomerase is shown to bind one of a pair of singly linked DNA rings (the DNA bearing the G-segment). Binding of ADPNP to the enzyme triggers the closure of the protein clamp and moves a T-segment through the G-segment. If the G- and T-segments reside on different rings, passage of one through the other leads to their decatenation. The one-gate model predicts that both rings remain trapped inside the closed enzyme clamp (*lower left*), whereas the two-gate model predicts that the ring bearing the G-segment remains trapped, but the ring bearing the T-segment can exit through the second protein gate and be released from the enzyme (*lower right*). See text and Roca, J. and Wang, J.C. 1994. *Cell* **77:** 609–616, for further details. (Illustration based on drawings kindly provided by R. Roca.)

SHARPENING THE IMAGE UNDER X-RAY

The two-gate model just described was based mainly on biochemical analysis. In 1994, three-dimensional structural information was available only for the N-gate, based on an X-ray diffraction study of crystals of an *E. coli* GyrB fragment containing the ATPase domain of the enzyme.[20] While biochemical studies increasingly favored the two-gate mechanism, X-ray diffraction studies of crystals of a yeast Top2 fragment were rapidly progressing.

In the mid-1990s, a fragment of the so-called B′–A′ region of yeast Top2 was found to form crystals suitable for X-ray diffraction studies. The B′–A′ region spans about 800 amino acid residues and contains the key parts of the enzyme for DNA breakage and rejoining. As is shown in Figure 5-10, the B′ portion corresponds to the carboxy-terminal half of the GyrB subunit (i.e., GyrB lacking its amino-terminal ATPase domain), and the A′ portion corresponds to the amino-terminal two-thirds of GyrA (i.e., that part of GyrA lacking its normal carboxy-terminal domain). By 1996, the structure of the

Even to this day, crystallography of large biological molecules is still an art—a trial-and-error game that often tests the patience and perseverance of a crystallographer. In the aqueous solution of a protein, interactions between the protein and solvent (mostly water molecules) tend to keep the protein molecules dispersed, and self-interactions between the protein molecules tend to congregate them; these two tendencies must be delicately balanced to form crystals in which the macromolecules are arrayed in a nicely ordered way; if not, the protein either remains dissolved or crashes out as an amorphous precipitate. This tug of war between two well-matched forces, one pulling macromolecules into solution and the other persuading them to form ordered arrays in a crystal, makes it very difficult —if not impossible—to predict (or even guess) which biological macromolecules can form well-ordered crystals and under what particular sets of conditions. The difficulty of persuading a protein to crystallize, and the number of trials involved in the crystallization of large biological molecules, explain why fragments rather than full-length proteins are often used in crystallographic studies: because the full-length proteins may have simply resisted forming crystals, or because they may yield crystals in which the molecules are not sufficiently ordered to produce adequate X-ray diffraction patterns.

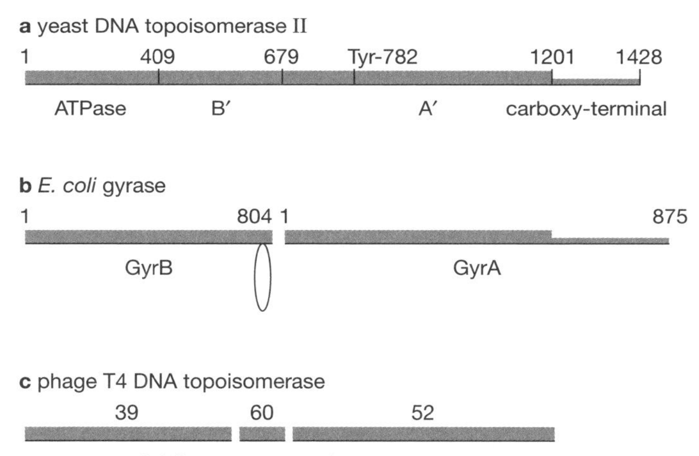

Figure 5-10. Comparison of different regions of type II topoisomerases. (a) *S. cerevisiae* DNA topoisomerase II is a single polypeptide 1428 amino acid residues in length. Three protease-sensitive sites around amino acid residues 409, 679, and 1201 divide the polypeptide into four regions, which are indicated as the ATPase, B′, A′, and the carboxy-terminal domains. Tyr-782 in the sketch marks the active-site tyrosyl residue. (*b*) *E. coli* gyrase is composed of two subunits GyrA and GyrB. The amino-terminal half of GyrB contains the ATPase region of the enzyme, and its amino acid sequence is homologous to the ATPase region of yeast Top2; the amino acid sequence of the carboxy-terminal half of GyrB is homologous to the yeast Top2 B′ region except for a region near the carboxyl terminus (the loop in the *middle* depiction). The amino-terminal two-thirds of the amino acid sequence of GyrA is homologous to the yeast A′ region. (There is little homology between the carboxy-terminal regions of yeast Top2 and *E. coli* GyrA.) (c) The phage T4 DNA topoisomerase is made up of three subunits, denoted by the genes encoding them. The carboxy-terminal region in the other two type IIA enzymes is absent in the phage topoisomerase. In the drawings, homologous and nonhomologous regions are represented by thicker and thinner lines, respectively. (Redrawn, with permission of Elsevier, from Berger, J.M. and Wang, J.C. 1996. *Curr. Opin. Struct. Biol.* **6:** 84–90.)

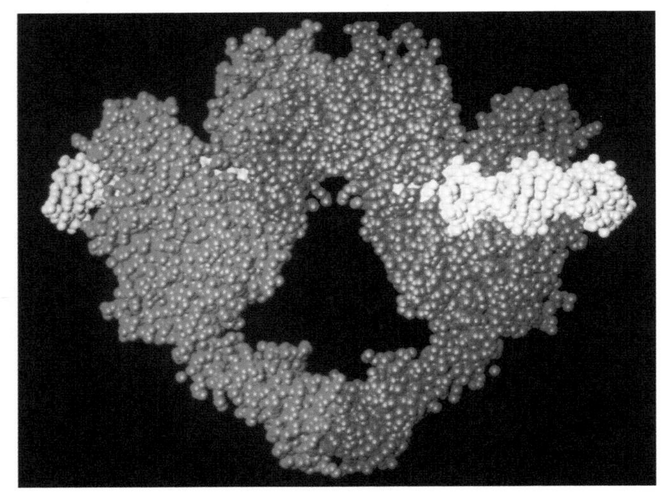

Figure 5-11. Structural model of the 800-amino-acid B′–A′ fragment of the dimeric yeast DNA topoisomerase II. The complex contains the two dimers (in red and blue) after its has cleaved a double-stranded DNA G-segment (yellow) and pulled the protein-linked broken DNA ends apart. The structure is depicted as a "space-filling" model in which the small spheres represent individual atoms (excluding the hydrogen atoms) in the macromolecules. See text for details. (Image based on the structure of the B′–A′ fragment reported by Berger, J.M., et al. 1996. *Nature* **379:** 225–232.)

B′–A′ fragment was solved.[25] Figure 5-11 illustrates a space-filling model of the structure of a dimer of the yeast Top2 B′–A′ fragment based on X-ray diffraction data, with a short piece of the G-segment modeled into the protein structure.

In all trials, no reasonable model could be built to accommodate an unbroken DNA G-segment in the protein structure. There is, however, a way to fit two broken halves of a DNA double helix into the protein structure. Each broken half has a short single-stranded overhang of four nucleotides with a 5′-phosphoryl end that extends from the enzyme-bound double-stranded region into a narrow tunnel within the protein structure, with the 5′-phosphoryl end of the overhang positioned next to the active-site tyrosyl residue of the enzyme (Tyr-782). All type IIA DNA topoisomerases are known to make a pair of staggered cuts, four base pairs apart, in the G-segment (Fig. 5-5). Thus, when the yeast enzyme cleaves a G-segment and opens the DNA gate, each 5′-phosphoryl end of the DNA is expected to join to a Tyr-782 residue, with four nucleotides extending from the protein-linked phosphoryl group to the double-stranded portion of the G-segment. The structural model depicted in Figure 5-11 is therefore consistent with that expected for an enzyme–DNA complex *after* unlocking and opening of the DNA gate in the G-segment. A prominent feature of the B′–A′ frag-

25. Berger, J.M., et al. 1996. *Nature* **379:** 225–232.

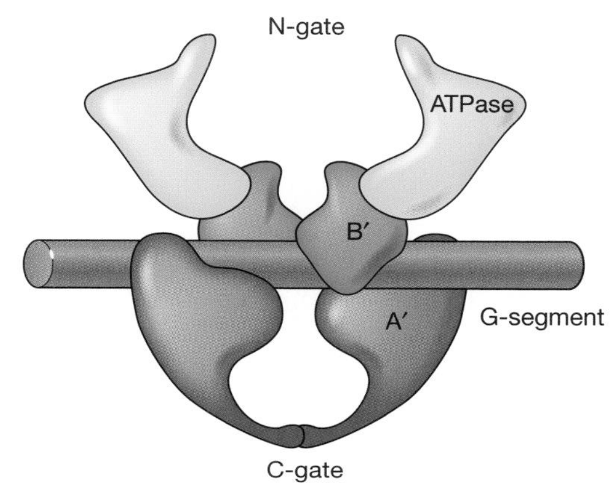

Figure 5-12. Schematic diagram of a type IIA DNA topoisomerase bound to a DNA G-segment based on the known structural features of the ATPase, B′, and A′ domains of type IIA DNA topoisomerases. The entrance and exit gates of the enzyme are marked as the N- and C-gate, respectively, and the DNA double helix is represented by a rod. See text for details. (Redrawn, with permission of the American Society for Biochemistry and Molecular Biology, from Olland, S. and Wang, J.C. 1999. *J. Biol. Chem.* **274:** 21688–21694.)

ment structure shown in Figure 5-11 is that the two halves of the dimer enclose a large hole, 55Å wide at its base, that can easily accommodate a DNA double helix with a diameter of about 20Å.

Together, the ATPase domain and the B′–A′ fragment constitute a core enzyme fully capable of catalyzing the ATP-dependent transport of one DNA double helix through another. Figure 5-12 depicts a model based on the three-dimensional structures of these parts available in the late 1990s, as well as the electron microscopic images of intact human and yeast DNA Top2 molecules.[26] Thus, the crystal structure of the B′–A′ fragment and that of the ATPase domain, determined initially for *E. coli* gyrase and later for the yeast enzyme, added many more details to the two-gate model first deduced from biochemical studies.

The G-segment-bound Top2 enzyme resembles a two-part contraption that functions like a Venus flytrap (Figs. 5-12 and 5-13). The two chambers are separated by the DNA G-segment; the "upper" chamber can open through the N-gate, to allow the T-segment to pass through this gate and enter the interior of the trap (Fig 5-13, left). (The DNA T-segment is likely attracted by the positively charged surface inside the N-gate.) When binding of ATP triggers closure of the N-gate, the T-segment is trapped, causing overcrowding in the upper chamber. The resulting strain in turn triggers an or-

26. Schultz, P., et al. 1996. *Proc. Natl. Acad. Sci.* **93:** 5936–5940; Benedetti, P., et al. 1997. *J. Biol. Chem.* **272:** 12132–12137.

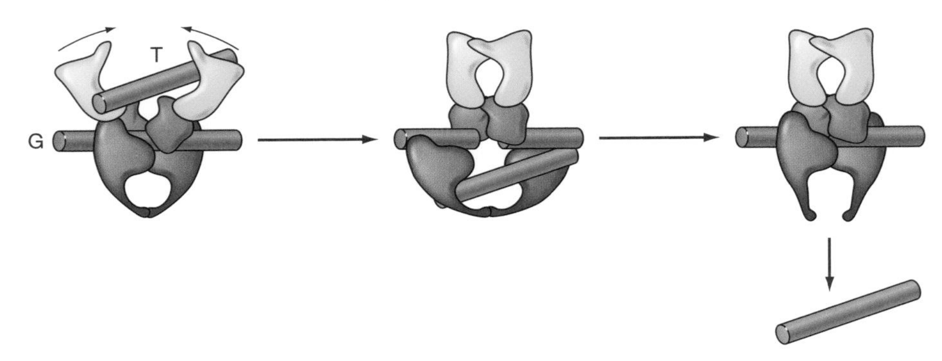

Figure 5-13. Transport of a DNA T-segment through an enzyme-bound DNA G-segment by a dimeric type IIA DNA topoisomerase. The T-segment DNA is shown entering through the N-gate (on the *left*), passing through the DNA gate of the G-segment (*middle* panel), and then leaving the complex through the exit gate of the enzyme (on the *right*). The ATPase, B′, and A′ domains are painted yellow, orange, and purple, respectively. The T-segment DNA is a green rod, the G-segment a blue rod. See text for details. (Redrawn, with permission of Macmillan Publishers Ltd., from Wang, J.C. 2002. *Nat. Rev. Mol. Cell Biol.* **3:** 430–441.)

derly conformation cascade in the enzyme–DNA complex. First, as the DNA gate in the G-segment is unlocked by the pair of active-site tyrosyl residues, the two enzyme halves come apart to open a 2.7-nm or 27Å gate in the DNA G-segment. The T-segment passes through this gate and enters the large hole in the lower chamber of the enzyme (Fig. 5-13, middle) that serves as a temporary quarter for the T-segment in transit. Once the T-segment has moved from the upper into the lower chamber, the upper chamber relaxes to resume its original structure, and the widely open DNA gate returns to its closed and locked position. Relaxation of the empty upper chamber, however, shrinks the size of the central hole in the lower chamber, and this contraction in turn creates overcrowding and strain. The strain within the lower chamber in turn favors expulsion of the T-segment through the exit-gate (Fig. 5-13, right). Finally, after the hydrolysis of the bound ATP molecules and release of ADP and P_i, the enzyme reopens its ATPase gate, ready for another cycle of DNA transport.[27,28]

27. Some of the mechanistic considerations are reviewed in Wang, J.C. 1998. *Q. Rev. Biophys.* **31:** 107–144.

28. The details of the conformational cascade, from the closure of the N-gate of a dimeric enzyme to the exit of the transported T-segment, are not yet known. There must be, however, intricate interplays among structural changes at various key locations in the enzyme–DNA complex, the hydrolysis of the two bound ATP molecules, and the release of the hydrolytic products. Coupling of the various conformational changes to ATP binding, hydrolysis, and release of the hydrolytic products is complex; there is good evidence, for example, that in the transport of one DNA through another by yeast Top2, two ATP molecules are sequentially hydrolyzed, and P_i release from hydrolysis of the first ATP probably precedes DNA breakage (Harkins, T.T. and Lindsley, J.E. 1998. *Biochemistry* **37:** 7292–7298; Harkins, T.T., et al. 1998. *Biochemistry* **37:** 7299–7312).

The three-dimensional structure shown in Figure 5-11 also suggested a more rigorous test of the two-gate mechanism. Inspection of the B′–A′ fragment structure suggested to us a way to lock the exit gate. If Lys-1126 (the lysine at position 1126 in the protein sequence) in one-half of the dimeric enzyme, and Asn-1042 (the asparagine at position 1042) in the other half are both replaced by cysteinyl residues, the sulfhydryl (–SH) groups of these latter residues would be positioned to form a disulfide (–S—S–) bond between them, upon oxidation by an exogenous oxidizing reagent (or by dissolved oxygen from the air). We made these amino acid changes, to create a K1126C–N1042C mutant Top2 (K, N, and C being, respectively, the one-letter amino acid codes for Lys, Asn, and Cys). Because yeast Top2 is dimeric, the mutant enzyme should have been capable of forming two symmetrically related pairs of cysteines across the putative exit gate. Formation of these disulfide bonds would lock the exit gate (Fig. 5-14), and prevent reopening of the gate by any subsequent conformational change in the Top2 enzyme. Thus, if the K1126C–N1042C mutant enzyme with its locked exit-gate were used in conjunction with ADPNP to mediate unlinking of a pair of singly interlocked DNA rings, then the DNA ring bearing the T-segment,

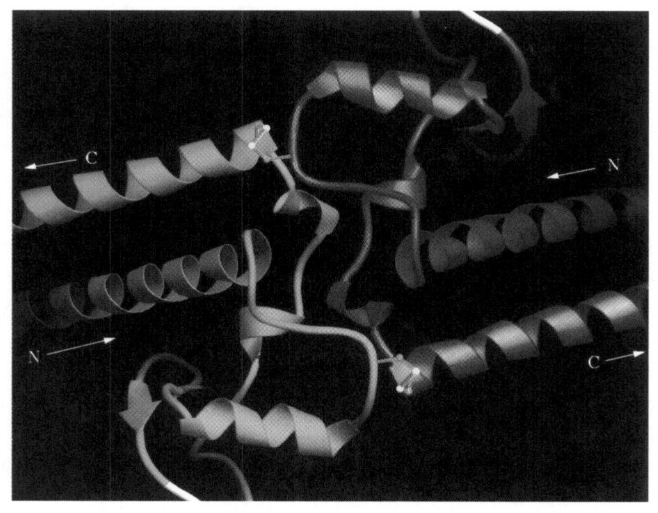

Figure 5-14. Locking the exit gate of yeast DNA topoisomerase II. The structure of the exit gate (C-gate) of *S. cerevisiae* Top2, viewed along the twofold symmetry axis of the enzyme in the direction of the entrance gate, is represented in a ribbon model. The two halves of the dimeric enzyme are colored purple and cyan, and the pair of disulfide bonds (yellow balls joined by orange sticks) joining the two pairs of cysteines in the N1042C–K1126C mutant yeast is shown. The two pairs of arrows in the sketch indicate a pair of antiparallel α-helices on either side of the C-gate. (The image is essentially the same as that shown in Roca, J., et al. 1996. *Proc. Natl. Acad. Sci.* **93:** 4057–4062, but adapted here with a pair of cysteines, rather than the original amino acid residues in the unmutagenized enzyme, forming the disulfide links.)

after its passage through the G-segment, would be unable to escape through the locked exit gate. Hence, both unlinked DNA rings would remain inside the protein ring, with its locked front and back doors.

We found this predicted outcome to be the case.[29] As a further step, we added a reducing reagent (2-mercaptoethanol) at the end of the unlinking reaction to reduce each disulfide bond of the mutant Top2 enzyme to two separate –SH groups. This treatment was expected to unlock the exit gate, and we did find that one of the pair of unlinked DNA rings within the closed protein clamp exited the protein ring while the other remained inside.[29] These experiments therefore showed unequivocally that the interface containing Lys-1126 and Asn-1042 did constitute an exit gate of the Top2 enzyme, and provided strong evidence for the two-gate mechanism of its action.

The preceding discussions of the general picture for how a type IIA DNA topoisomerase works omitted many of the structural details of various conformational changes, and the rather complex coupling between various conformational changes and ATP binding, hydrolysis, and the release of the hydrolytic products of ATP.[28] Many of these aspects have yet to be determined by biochemical, structural, and kinetic studies of these fascinating enzymes. Some of the conjectures may turn out to be off the mark, whereas others are most likely correct. In the examples shown in Figures 5-12 and 5-13, the location of the DNA G-segment in the B′–A′ core structure, and the formation of an exit gate by a pair of symmetry-related domains near the carboxy-terminus of A′, have been further supported by a subsequently determined structure of the yeast Top2 B′–A′ fragment noncovalently bound to a short double-stranded DNA.[30]

A BLACK SHEEP IN THE FLOCK

The preceding sections summarize key experiments that led to the two-gate protein-clamp model for a type IIA DNA topoisomerase. One question remains, however, about why gyrase is so different among members of the type IIA subfamily. Why is gyrase the only member that catalyzes the negative supercoiling of a relaxed DNA ring, whereas all the others catalyze the relaxation of negatively or positively supercoiled DNA rings?

To address this question, it is useful to consider DNA transport by a type IIA enzyme as occurring in two stages: First, a T-segment enters the N-gate of an enzyme bound to a G-segment. Second, upon closure of the N-gate following the binding of ATP, the T-segment that has entered is sequentially moved through the DNA gate and the protein exit gate. It is likely that the nuances of DNA transport by the different type IIA enzymes are manifestations of the first stage. We have long suspected that the unique ability of gyrase to supercoil a DNA is closely tied to the way the enzyme interacts with a pair of G- and

29. Roca, J., et al. 1996. *Proc. Natl. Acad. Sci.* **93:** 4057–4062.

30. Dong, K.C. and Berger, J.M. 2007. *Nature* **450:** 1201–1205. Amazingly, in this enzyme–DNA co-crystal structure, the exit gate of the enzyme is open, which adds further evidence for the two-gate model.

T-segments. For other type IIA DNA topoisomerases, a DNA-bound dimeric enzyme shields a 25-bp region of the DNA from nuclease cleavage.[31] The sites at which the topoisomerase itself transiently cleaves the DNA are located near the center of this 25-bp region, thus corresponding to the enzyme-shielded G-segment shown in the enzyme–DNA model depicted in Figure 5-12. In the case of bacterial gyrase, however, enzyme binding shields a much longer DNA segment: A 140-bp DNA segment is protected from digestion by staphylococcal nuclease.[23] Probing with pancreatic DNase I showed a nearly refractive region of ~30 bp in the midsection of this 140-bp segment, whereas in the two flanking regions of this segment the nuclease cleaves with a characteristic spacing of 10 or 11 nucleotides between adjacent sites on the same DNA strand.[23,32] This 10- or 11-nucleotide spacing had been seen before: It is related to the helical periodicity of DNA in solution and is indicative of a DNA double helix lying on a smooth surface, and DNA adsorbed to calcium phosphate crystals or a mica surface also shows such a periodic DNase I cleavage pattern.[23,33,34] Thus gyrase is unique among the type IIA enzymes in that each half of the dimeric enzyme has an additional DNA-binding domain for wrapping, onto its surface, a DNA segment ~50 bp long.

Wrapping of the flanking regions of a 140-bp DNA segment around gyrase apparently imposes a right-handed writhe in the DNA.[35] As far back as 1978, we observed that the binding of increasing numbers of gyrase molecules to a nicked DNA ring in the absence of ATP yielded DNA rings with increasing values of *Lk* upon sealing of the nicks with *E. coli* DNA ligase (recall that this enzyme requires NAD but not ATP, and thus can seal the nicks in the absence of ATP).[35] There is strong evidence that the pair of carboxy-terminal domains of the GyrA protomers is involved in wrapping the flanking regions of a 140-bp G-segment around the gyrase. First, *E. coli* GyrA (572–875), a fragment containing the carboxy-terminal 304 amino acid residues of GyrA, has been shown to bind DNA and induce a positive writhe.[36] Second, the crystal structure of the GyrA carboxy-terminal domain suggests that it has a distribution of surface charges suitable for wrapping a negatively charged DNA double helix around it.[37] Third, gyrase lacking the carboxy-terminal domain of GyrA can no longer impose a right-handed writhe in an enzyme-bound DNA, and furthermore, the deletion of this domain abolishes the abil-

31. See, for example, Lee, M.P., et al. 1989. *J. Biol. Chem.* **264:** 21779–21787.

32. Kirkegaard, K. and Wang, J.C. 1981. *Cell* **23:** 721–729; Morrison, A. and Cozzarelli, N.R. 1981. *Proc. Natl. Acad. Sci.* **78:** 1416–1420; Fisher, L.M., et al. 1981. *Proc. Natl. Acad. Sci.* **78:** 4165–4169.

33. Rhodes, D. and Klug, A. 1980. *Nature* **286:** 573–578.

34. The 10–11 nucleotide spacing of DNase I cleavage sites was first seen in studies of nucleosomes, in which DNA is wrapped around a protein core made of four different histones (see Whitlock, J.P. Jr. and Simpson, R.T. 1977. *J. Biol. Chem.* **252:** 6516–6520; Lutter, L.C. 1978. *J. Mol. Biol.* **124:** 391–420).

35. Liu, L.F. and Wang, J.C. *1978. Proc. Natl. Acad. Sci.* **74:** 2098–2102.

36. Reece, R.J. and Maxwell, A. 1991. *Nucleic Acids Res.* **19:** 1399–1405.

37. Corbett, K.D., et al. 2004. *Proc. Natl. Acad. Sci.* **101:** 7293–7298; Ruthenburg, A.J., et al. 2005. *J. Biol. Chem.* **280:** 26177–26184.

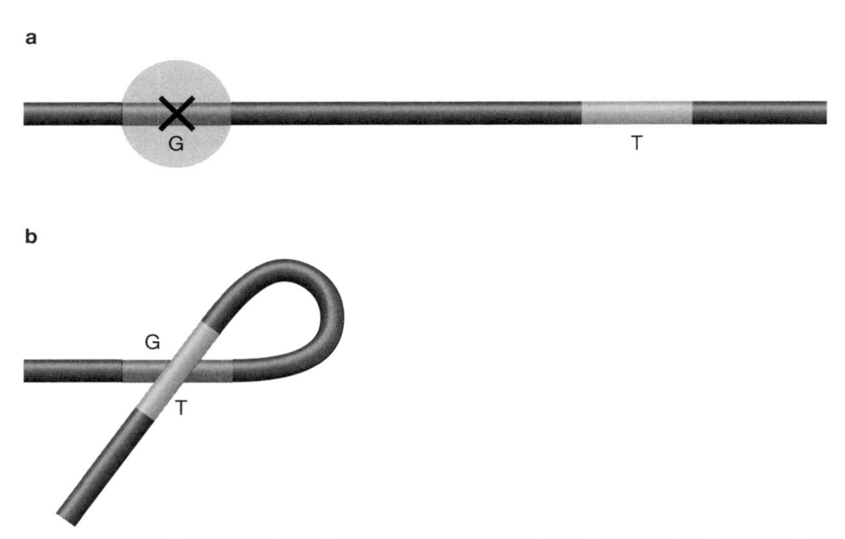

Figure 5-15. Curving of a gyrase-bound DNA G-segment and its effect on the directional transport of a DNA T-segment on the same DNA through the G-segment. (*a*) The long rod represents a DNA bearing a G-segment and a T-segment; the circle represents a G-segment- bound gyrase. The cross at the center of the circle marks the position of the molecular dyad of the DNA-bound enzyme, which passes through the complex in the direction from the entrance gate to the exit gate. (*b*) The right-handed wrapping of a DNA segment around a gyrase affects its intramolecular transport of a T-segment through a G-segment in a directional way. See text for details. (Redrawn and modified from Wang, J.C. 1998. *Q. Rev. Biophys.* **31:** 107–144.)

ity of gyrase to negative supercoil a DNA ring.[38] The truncated enzyme behaves just like all other type IIA enzymes, catalyzing the ATP-dependent relaxation of a positively or negatively supercoiled DNA but not the negative supercoiling of a relaxed DNA ring.[38]

This dependence of the supercoiling activity of gyrase on the presence of the GyrA carboxy-terminal domain provides strong support for the notion that the unique catalytic characteristics of gyrase are manifestations of its interaction with the G-segment. Figure 5-15a shows a type IIA enzyme bound to the G-segment, in which the enzyme is viewed down its molecular dyad. In the case of gyrase, the right-handed wrapping of a DNA segment around it is likely to have two effects: First, this wrapping tends to bring a DNA segment juxtaposed to the G-segment to the interior of the N-gate when it is open, thus making this juxtaposed segment the T-segment (Fig. 5-15b). In other words, the right-handed wrapping of a flanking region of the G-segment favors the selection of a T-segment contiguous with the G-segment. Thus, gyrase is expected to favor an intramolecular transport of one DNA segment through another,

38. Kampranis, S.C. and Maxwell, A. 1996. *Proc. Natl. Acad. Sci.* **93:** 14416–14421.

which would generally lead to a change in supercoiling, rather than intermolecular DNA transport, which would lead to catenation or decatenation. Second, the right-handed wrapping imposes a positive node between the T-segment and the enzyme-bound G-segment (the nodal sign is chosen so that a positive node corresponds to the kind of crossover seen in a positive supercoiled DNA). ATP binding then triggers closure of the N-gate and transport of the contiguous T-segment through the DNA gate.

Therefore it appears that gyrase can negatively supercoil a DNA ring because its transport of DNA is directional: It imposes a positive node between a T-segment to be transported and the G-segment, and then inverts this node to a negative one. Prior to DNA transport, it is only in the case of gyrase that the nodal sign between the G- and T-segment is specified by the enzyme; in all other cases, the DNA specifies the sign—positive nodes in a positively supercoiled DNA and negative nodes in a negatively supercoiled DNA—and DNA transport always reduces the net number of nodes needed to relax the supercoiled DNA.[27]

A DISTINCT BRANCH OF THE TYPE II DNA TOPOISOMERASE FAMILY: THE TYPE IIB ENZYMES

For two decades, it was thought that all type II DNA topoisomerases constituted a single subfamily that were structurally and mechanistically very closely related. In 1997, however, a type II enzyme purified from *S. shibatae*, the same sulfur-loving hot-spring microorganism from which the first reverse gyrase was discovered (see Chapter 3), was found to differ significantly from the other known type II DNA topoisomerases. Like the known bacterial type II DNA topoisomerases gyrase and DNA topoisomerase IV, the *S. shibatae* enzyme has two subunits, A and B, and utilizes ATP in its relaxation of supercoiled DNA or unlinking of catenated DNA rings. However, the amino acid sequences of the *S. shibatae* enzyme subunits revealed little resemblance to those of the other type II enzymes, except for the presence of three motifs within the putative ATPase domain of its B subunit. Thus, the *S. shibatae* enzyme, named DNA topoisomerase VI in keeping with the tradition of using even Roman numerals to designate the type II enzymes, appeared to represent a new kind of type II DNA topoisomerase. The type II enzymes were therefore divided into two subtypes, IIA and IIB. The new enzyme found in the archaeon *S. shibatae* became the founding member of the type IIB subfamily, whereas the type II enzymes of the *E. coli* gyrase and yeast DNA topoisomerase II category were designated the type IIA subfamily.[13,17]

Enzymes similar to *S. shibatae* DNA topoisomerase VI were soon identified in many other archaea, as well as in plants like the weed *Arabidopsis thaliana*.[39] An-

39. Yin, Y., et al. 2002. *Proc. Natl. Acad. Sci.* **99:** 10191–10196; Hartung, F., et al. 2002. *Curr. Biol.* **12:** 1787–1791. For short reviews of type IIB DNA topoisomerases in plants, see Gadelle, D., et al. 2003. *BioEssays* **25:** 232–242; Corbett, K.D. and Berger, J.M. 2003. *Chem. Biol.* **10:** 107–111.

other protein that shared amino acid sequence similarities with the A subunit of DNA topoisomerase VI, the yeast protein Spo11, was found to initiate meiotic recombination by cleaving double-stranded DNA and simultaneously forming covalent protein–DNA links.[13,40] Meiosis is the process of germ cell formation in sexually reproducing organisms, and meiotic recombination is the process by which a pair of matching paternal and maternal chromosomes exchange genetic material at a small number of more or less randomly occurring locations along them. Meiotic recombination is important for the proper sorting of paired chromosomes and also enhances genetic variations in the gametes so produced. Yeast Spo11 turns out to be a member of a family of proteins present in diverse organisms including fruit flies, worms, plants, and mammals. Its initiation of meiotic recombination by double-stranded DNA cleavage through a transesterification mechanism smacks of a type II DNA topoisomerase: however, the Spo11 protein does not appear to participate directly in the rejoining of the broken DNA ends it creates. Instead, before joining of the DNA ends by a meiotic recombination system, the Spo11 protein is removed by a nuclease activity that cuts off a short piece of the DNA strand covalently linked to it.[41]

STRUCTURAL SIMILARITIES BETWEEN THE TYPE IIA AND IIB DNA TOPOISOMERASES

In 1999, it became clear that the type IIB DNA topoisomerases are more closely related to the type IIA enzymes than the comparison of the amino acid sequences of these enzymes, done 3 yr earlier, had suggested. New evidence came from resolution of the crystal structure of a large fragment of DNA topoisomerase VI of the methane-producing bacterium *Methanococcus jannaschii*, an amazing organism that lives at high temperatures near hydrothermal vents on the deep sea floor. This structure revealed that the two motifs important in DNA binding and cleavage by the type IIA DNA topoisomerases are also present in the type IIB *M. jannaschii* enzyme, even though the amino acid sequences of these motifs in the IIA and IIB enzymes had diverged sufficiently to obscure any sequence similarity.[42] The relatedness between the two type II subfamilies of enzymes became even more evident with the solution of the X-ray structure of the *S. shibatae* DNA topoisomerase VI B subunit containing the ATPase domain of the enzyme.[43] Whereas previously only three short motifs within the ATPase domains of the type IIA and IIB

40. Keeney, S., et al. 1997. *Cell* **88:** 375–384.

41. Neale, M.J., et al. 2005. *Nature* **436:** 1053–1057.

42. Nichols, M.D., et al. 1999. *EMBO J.* **18:** 6177–6188.

43. Corbett, K.D. and Berger, J.M. 2003. *EMBO J.* **22:** 151–163.

DNA topoisomerases were known to share amino acid sequence homology, the structure of the archaeal enzyme B subunit shows that nearly the entire subunit is folded into a three-dimensional structure resembling the type IIA DNA topoisomerase ATPase domain.[43,44]

Nevertheless, the division of the type II DNA topoisomerases into two subfamilies, IIA and IIB, is fully justified, because there are major differences in the architectures of the two subfamilies of proteins. Figure 5-16 depicts a crystal structure of a full-length DNA topoisomerase VI from *Methanosarcina mazei*, another methane-producing archaeon often found in wetland mud.[45] In the B_2A_2 or $(BA)_2$ structure of the whole IIB enzyme (a IIB "holoenzyme"), the two B protomers form an ATP-modulated gate similar to the N-gate of the type IIA enzymes. But the large central hole in the type IIA enzyme (shown in Figs. 5-11 and 5-12), bounded on its top by domains holding the G-segment and on its bottom by domains forming the C-gate, is conspicuously missing from the structure shown in Figure 5-16. The structures of both the DNA-binding core of *M. jannaschii* DNA topoisomerase VI and the *M. mazei* DNA topoisomerase VI holoenzyme have a deep groove across the pair of A protomers, and a DNA G-segment can be modeled into this groove.[42,45] Whereas a type IIA enzyme has a distinct C-gate, well separated from the G-segment for the exit of the T-segment after its passage through the DNA, in a type IIB enzyme the C-gate of the protein and the DNA gate have coalesced into a single exit gate. Because the crystal structures of the type IIB enzymes show that there is simply no room to moor a T-segment after it has sailed through a transiently opened DNA gate, the DNA gate and the protein exit gate in a type IIB enzyme most likely operate in unison as a single structural unit in an enzyme–DNA complex.

The major steps proposed for DNA transport by a type IIB DNA topoisomerase otherwise generally resemble those postulated for the type IIA enzymes. As with the type IIA DNA topoisomerases, the various steps in DNA transport by a type IIB enzyme are intricately coupled to the binding and hydrolysis of ATP, as well as to the release of the hydrolytic products of ATP. Several crystal structures of the ATPase domain of *S. shibatae* DNA topoisomerase VI, by itself and in complexes with compounds mim-

44. This structure is also homologous to the ATPase domains of several other proteins, including MutL (see Footnote 20 above), and Hsp90, a ubiquitous protein in bacteria and eukarya. Hsp stands for "heat shock protein," and is so named because the cellular concentration of the 90-kDa protein is increased when cells are subject to stresses such as a sudden increase in temperature. A limited homology among the ATPase domains in these proteins was first recognized from a comparison of their amino acid sequences (see Reece, R.J. and Maxwell, A. 1991. *Nucleic Acids Res.* **19:** 1399–1405). The high degree of structural homology among the ATPase domains of these proteins was also seen in the crystal structures of MutL and Hsp90.

45. Corbett, K.D., et al. 2007. *Nat. Struct. Mol. Biol.* **14:** 611–619. The structure was resolved to a resolution of 4Å. At such a resolution, structural features such as the α-helices and β-sheets are readily recognizable in the polypeptide chain, but the positions of the individual atoms are not discernible. The individual amino acid residues can be positioned along the path of the polypeptide in the crystal, however, owing to the known amino acid sequence of the polypeptide.

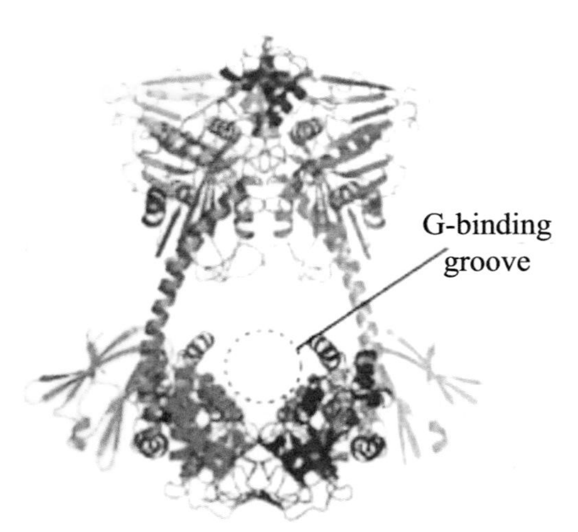

Figure 5-16. Structural model of a full-length type IIB DNA topoisomerase of the archaeon *Methanosaurcina mazei*. The crystal structure depicted is viewed along a likely binding cleft for the DNA G-segment. The two ATPase domains forming the entrance gate are in contact with each other despite the absence of ATP or its nonhydrolyzable analog ADPNP in the crystallization mixture, but the dimer contacts are not as extensive as those observed in a pair of ADPNP-bound type IIB enzyme ATPase domains. One AB dimer (the one on the *right*) is shown in gray; various domains of the other dimer are differently colored (the ATP-binding domain in yellow, the A subunit domains in green and blue, and the transducer domain in orange). (Reprinted, with permission of Macmillan Publishers Ltd., from Corbett, K.D., et al. 2007. *Nat. Struct. Mol. Biol.* **14:** 611–619.)

icking different states of ATP during and after its hydrolysis,[46] indicate that ATP binding to a protomer pair facilitates their dimerization, as first revealed by biochemical and structural studies of the type IIA enzymes. ATP binding appears to change the conformation of the type IIB enzyme to an "ATP-restrained" state. In this state, the amino acid residues in the ATPase domain for catalysis of ATP hydrolysis are properly poised, and there is also a relative rotation of about 11°, from their positions in the ATP-free state, between the ATPase domain and a "transducer domain" (see Fig. 5-16). The transducer domain connects the B subunit ATPase domain to the A subunit DNA breakage and rejoining domain through a long α-helix, and the 11° rotation is presumably a key step in transmitting the ATP-actuated conformation change within the B subunit to the A subunit for activating the exit of the entrapped T-segment.[43,46] The restrained state persists during ATP hydrolysis, and reversion to the relaxed state occurs only after the DNA passage event to permit the release of the sequestered ATP hydrolytic products.

46. Corbett, K.D. and Berger, J.M. 2005. *Structure* **13:** 873–882.

KINSHIP BETWEEN THE TYPE IA AND TYPE IIA DNA TOPOISOMERASES

The inventions of Nature can be dazzling and bewildering at times. Yet, at least during the billions of years of life on Earth, Nature has also shown a remarkable tendency to stick with a good invention, making gradual adjustments rather than fundamental design changes. This discussion of the type IIA and IIB DNA topoisomerases serves as a good example.

At first glance, the pictures presented in Chapters 3 and in this chapter for the type IA and type II enzymes appear very different. One is a single polypeptide that breaks one DNA strand at a time to allow the passage of another strand (or perhaps of a DNA double helix); the other consists of two identical halves and transiently breaks both strands of a DNA double helix to allow the passage of a second DNA double helix. One is in the form of a C-clamp with a single protein gate for the entrance as well as the exit of the passing DNA strand; the other is a DNA transport machine with separate ports for the entrance and exit of the DNA double helix being transported. One has no requirement for ATP; the other has a strict dependence on ATP.

Yet a closer look at the three-dimensional structures of the type IA and type II enzymes reveals some common shared features: The two groups of enzymes probably use very similar structural motifs in their breakage and rejoining of DNA strands. In all type IA and type II enzymes, the active-site tyrosyl residue is perched on an extended loop in a structurally similar scaffold, and several other residues that are most likely involved in DNA breakage and rejoining reactions are also positioned in nearly identical ways in a structurally similar fold.[47] These common structural features clearly indicate a kinship among type IA and type II DNA topoisomerases. Thus, three of the four subfamilies of the DNA topoisomerases: IA, IIA, and IIB, appear to be structurally and evolutionarily related. Only the fourth subfamily, the type IB subfamily, appears to have a different lineage. But why is there a need for all of these DNA topoisomerases? Would one not be sufficient in solving the DNA untanglement problem? These are the questions that we will address in Chapters 7 and 8.

47. Berger, J.M., et al. 1998. *Proc. Natl. Acad. Sci.* **95:** 7876–7881. The fold containing several catalytically important acidic amino acid residuals found in both type IA and IIA DNA topoisomerases, initially termed a "Rosmann-like fold," is now commonly referred to as a "toprim" fold because of its presence in many proteins including the DNA topoisomerases and primases that prime DNA synthesis in cells (see Aravind, L., et al. 1998. *Nucleic Acids Res.* **26:** 4205–4213).

CHAPTER 6

Real-Time Viewing of the DNA Topoisomerases in Action

"You can see a lot by watching."
Yogi Berra

THE PRECEDING THREE CHAPTERS DESCRIBE HOW different subfamilies of DNA topoisomerases perform their magic. Because the actions of these enzymes can be followed readily by monitoring topological changes in DNA rings, it was the discovery of DNA rings, especially supercoiled DNA rings, that set the stage for the discovery of DNA topoisomerases (as we saw in Chapters 1 and 2). Ring-shaped DNA was also used almost exclusively in early experiments probing the catalytic properties of these enzymes.

Since 1990, however, new methods that do not rely on the use of DNA rings have been developed to study DNA topology and DNA topoisomerases. Several of these methods are based on the use of micromanipulators, such as the magnetic and optical "tweezers" or "traps," whereas others rely on observation of the molecules without manipulation. In an experiment involving micromanipulation, one end of a DNA molecule in a thin layer of solution is rigidly anchored to a solid surface or the shaft of a computer-controlled motor, and the other end to a micrometer (μm)-sized bead that can be moved about by magnetic forces or light beams. By tracking the positions of the bead through the objective lens of a high-sensitivity video camera, the spatial coiling of the bead-attached DNA and the forces it is subject to can be monitored in real time. The action of a DNA topoisomerase on a bead-linked DNA, which can alter its spatial coiling, can be similarly monitored.

In this type of experiment, individual macromolecules are examined one by one. The DNA molecules with or without bound proteins can be individually manipulated and observed in real time, under physiological conditions, during the course of an experiment. As the next several sections will show, these unique aspects make the use of micromanipulators particularly advantageous in the study of DNA topoisomerase actions.

Not all methods examining individual biological macromolecules involve the use of micromanipulators, however. Electron microscopy has long been used to view individual macromolecules, and we have seen several examples of its use in the preceding chapters. Before the development of what is termed "cryoelectron microscopy," however, viewing of biological macromolecules required placing them in a high vacuum, owing to the strong scattering of electrons by all molecules in the path of an electron beam. This exposure to vacuum in turn required that the macromolecules be first adhered to a supporting surface and dried, before their viewing in the electron beam. Furthermore, to enhance image contrast, it was usually necessary to stain the macromolecules with a reagent that strongly scatters electrons, or coat them with a very thin metal film. Biological macromolecules viewed under these conditions are therefore far from their native environment, and their biological activities are rarely preserved.

With the development of cryoelectron microscopy, however, harsh treatments of the biological macromolecules before viewing in an electron beam are largely avoided. In this method, a thin layer of a solution containing the macromolecules is flash-cooled, so as to rapidly form a sheath of amorphously frozen water around the macromolecules. The sample holder containing this frozen sheet is also kept at very low temperature to maintain the vapor pressure of ice to a tolerable level during direct viewing of the frozen sheet in an electron beam. Adherence of the macromolecules to a supporting film is not necessary because the frozen sheet itself provides a supporting matrix for the embedded macromolecules. Because the electrons are more strongly scattered by the embedded macromolecules than by their surrounding water molecules, adequate contrast of the imaged macromolecules can usually be achieved without contrast enhancement, especially when computational methods are used that superimpose the images of many individual macromolecules to improve the signal-to-noise ratio. In addition, rapid freezing also provides a way to catch the macromolecules in various conformations in solution. All these advantages make cryoelectron microscopy one of the most powerful techniques in studying the structures and dynamic aspects of biological macromolecules.

Another approach for studying individual macromolecules is atomic force microscopy. In this technique, a sharp nanometer (nm)-sized tip is used to scan a thin layer of solution spotted on a solid surface. The height of the tip above the surface and its position in the plane parallel to the surface are controlled and varied by a computer. As the probe descends at a particular position in the horizontal plane to approach a macromolecule in solution, the force sensed by the probe is monitored. The variations observed in this force at different heights and positions of the tip provide information for contouring the macromolecules within the thin layer of solution.

In addition to these methods, fluorescence microscopy is also widely used in studying macromolecules at the single-molecule level. In 1983, the Brownian or thermal movements of individual DNA molecules, which were made fluorescent by binding of ethidium molecules (Chapter 3), were first visualized and studied by the use of an optical microscope equipped with a highly sensitive video camera.[1] Individual DNA molecules migrating through a gel matrix under the influence of an electric field were similarly viewed and studied, and the twisting and turning of the thread-like molecules during their fervent march toward the positive electrode were monitored in real time.[2]

Perhaps the most powerful approach in applying fluorescence spectroscopy to the study of DNA topoisomerase actions is Förster resonance energy transfer (FRET), also termed fluorescence resonance energy transfer. FRET is one particular mechanism by which a donor fluorophore (D) dissipates the excess energy it acquires when one of its electrons is excited to a higher energy level by the absorption of a photon (or, alternatively, in the case of chemiluminescence, owing to a chemical reaction). In general, a part of this excitation energy is rapidly dissipated through thermal motions of atoms within the macromolecule, or through collisions with the surrounding small molecules. The remaining excitation energy can then be released by emitting a photon (donor fluorescence) or transferred to a second (acceptor) fluorophore A if one with suitable spectral characteristics is nearby. In the latter case, the excited A can in turn emit a photon (acceptor fluorescence). There are several mechanisms of transferring excitation energy from D to A, among which FRET is probably the most important one in biological studies.

FRET occurs through space or any medium between D and A, and its efficiency is inversely proportional to the sixth power of the distance between D and A. It is this steep dependence on distance that makes FRET a powerful molecular ruler. A number of spectroscopic measurements can be used to measure FRET efficiency: for example, by measuring a decrease in donor fluorescence owing to the presence of the acceptor or an increase in acceptor fluorescence owing to the presence of the donor. Although FRET has long been used to measure molecular distances, it is only in recent years that improvements in detector sensitivity have made it feasible to perform such measurements at the single-molecule level.[3] Detector sensitivity and noise problems impose an upper limit of about 10 nm for applying FRET to distance measurements, and there is also a lower limit of the D–A separation, about 1 nm, below which excitation energy transfer by a mode different from FRET becomes significant. Because the actions of various DNA topoisomerases often involve movements in the 1–10-nm range between different parts in an enzyme–DNA complex, FRET measurements provides an ideal method for observing such movements.

1. Yanagida, M., et al. 1983. *Cold Spring Harb. Symp. Quant. Biol.* **47:** 177–187.

2. Schwartz, D.C. and Koval, M. 1989. *Nature* **338:** 520–522.

3. For representative reviews, see Weiss, S. 1999. *Science* **283:** 1676–1683; 2000. *Nat. Struct. Biol.* **7:** 724–729; Heyduk, T. 2002. *Curr. Opin. Biotechnol.* **13:** 292–296; Johnson, C.K., et al. 2005. *Physiology* **20:** 10–14; Greenleaf, W.J., et al. 2007. *Annu. Rev. Biophys. Biomol. Struct.* **36:** 171–190.

In this chapter, we shall first consider magnetic and optical traps and end with a discussion of an experiment using single-molecule FRET. In this last example, the action of a DNA topoisomerase is deduced from direct observation of a change in FRET between two fluorophores placed at specific locations in a DNA, and so involves neither the use of a micromanipulator nor the attachment of beads to DNA ends.

MOLECULAR GLUES

Before a micromanipulator can be used to coil up a DNA molecule, a micron-sized bead is often attached to one or both ends of the DNA, and one DNA end is often anchored to a solid surface, such as the inner wall of a glass capillary tube. Specific "molecular glues" have been developed for attaching DNA molecules to beads or other surfaces; typically these are pairs of components that interact strongly and specifically with each other. One of the favorite glues uses biotin, a small vitamin, and avidin, a protein in egg white that binds biotin very tightly (a bacterial protein streptavidin can be used instead of avidin). Another popular molecular glue uses antibodies that bind specifically and tightly to a plant steroid called digoxigenin.

To apply these molecular glues to an end or a particular region of a DNA, suitable nucleoside derivatives have been synthesized in which either a biotin or a digoxigenin moiety is chemically tethered to a DNA base. Among these, cytosine derivatives with their C5 position modified are often used for incorporation into a DNA oligomer, which can be subsequently joined to a particular end of a DNA, or for enzymatic incorporation into a specific region of a long DNA. As examples, biotin- or digoxigenin-substituted oligonucleotides, each with a nucleotide sequence tailor-made for joining to one particular end of a linear DNA, can be used to specifically anchor the two ends of a DNA molecule to two different surfaces, one coated with avidin and the other with specific antibodies against digoxigenin. In general, the higher the number of biotin- or digoxigenin-substituted nucleotides at an end or a particular internal location of a DNA, the stronger the bonding between the DNA and the appropriately coated solid surface.

MAGNETIC TRAP[4]

Magnetic traps or tweezers are particularly convenient for examining supercoiling or other topological changes in individual DNA molecules. In the arrangement shown in Figure 6-1a, a glass capillary or microchamber is placed on the viewing stage of an inverted microscope. One end of a DNA molecule is fixed to the surface of the capillary or microchamber, and the other end is attached to a micron-

4. Bustamante, C., et al. 2000. *Nat. Rev. Mol. Cell Biol.* **1:** 130–136; Charvin, G., et al. 2005. *Annu. Rev. Biophys. Biomol. Struct.* **34:** 201–219; Zlatanova, J. and Leuba, S. H. 2003. *Biochem. Cell. Biol.* **81:** 151–159.

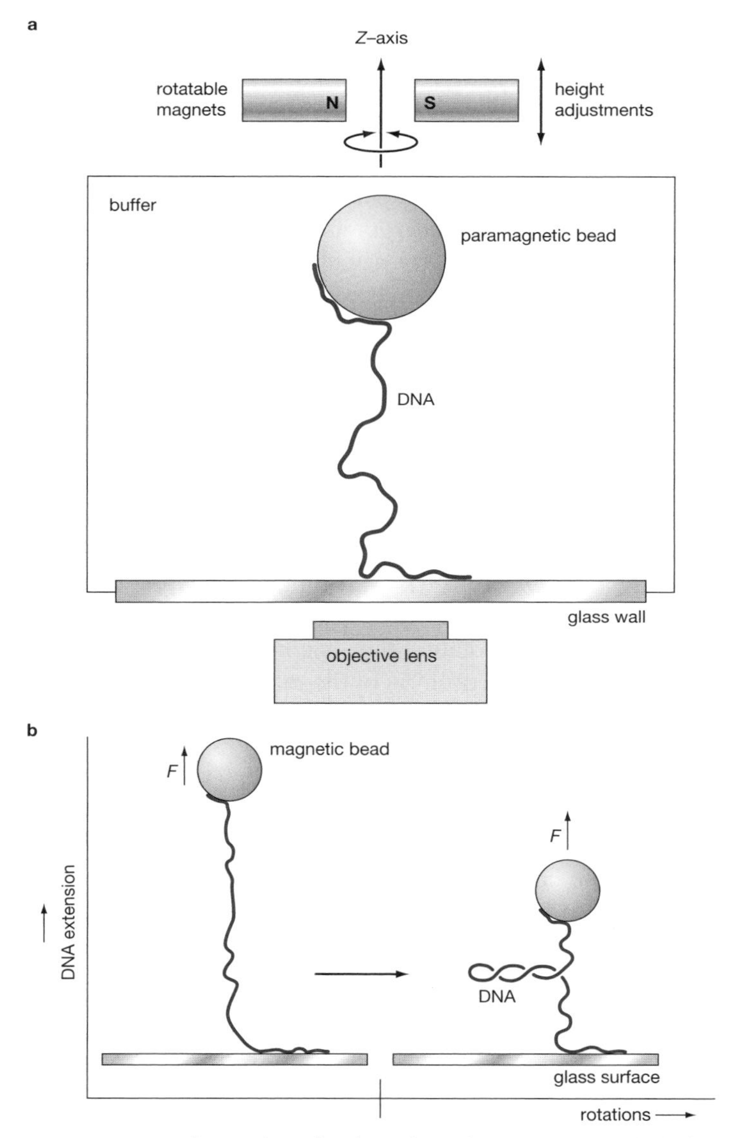

Figure 6-1. A magnetic trap used in single-molecule analysis of DNA. (*a*) One end of a linear DNA molecule is attached to a micron-sized paramagnetic bead, and the other end is affixed to the wall of a glass capillary or microchamber containing an appropriate buffer. A pair of permanent magnets above the sample chamber is used to hold the magnetic bead up in the vertical or *z*-direction. The position of the bead is monitored through the objective lens of an inverted microscope. Different parts in the illustration are not drawn to scale. (*b*) The extended DNA can be twisted by rotating the magnets around the *z*-axis. After a certain number of turns, the DNA buckles under torsion to form a supercoiled loop, which in turn leads to a decrease in the vertical extension $<z>$ of the DNA. (Redrawn, with permission of Elsevier, from Strick, T., et al. 2000. *Prog. Biophys. Mol. Biol.* **74:** 115–140.)

sized paramagnetic bead. A small permanent magnet mounted on an adjustable stage is placed over the sample chamber to pull the bead upward. The pulling force *F* that stretches the DNA can be controlled by adjusting the vertical position of the magnet, and its magnitude can be calculated by a number of methods. Because the force stretching the DNA has a direct effect on the thermal motions of the DNA-attached bead in the horizontal plane—the greater the force, the more difficult it is for the suspended bead to wander from its equilibrium position in the horizontal plane—one convenient method for calculating this force is from the observed amplitude of the thermal fluctuations in the bead position in the horizontal plane.[5]

For a paramagnetic bead, its magnetic dipole tends to align with the direction of the magnetic field of the permanent magnet. If this aligning torque is sufficiently strong, and if both strands at each DNA end are tightly glued to the surface the end is anchored to, then the DNA between the bead and the other anchored end constitutes a topological domain, with a well-defined linking number *Lk* between its two intertwined DNA strands. Unlike the case of a DNA ring, however, *Lk* can now be increased or decreased by a particular number by simply rotating the pair of permanent magnets a corresponding number of right- or left-handed turns. In the case of an individual DNA ring *Lk* must be an integer, whereas here no such restriction exists because the two ends are not joined to each other.

The ability to alter *Lk* by rotating the magnet, as well as to alter the pulling force *F* by adjusting the vertical position of the magnet, makes the magnetic trap a powerful tool in studying the energetics of DNA supercoiling.[4,5] At a given pulling force *F*, a supercoiled loop begins to form in the DNA after a certain number of rotations of the bead held in the magnetic trap. The exact number of rotations for this "buckling" of the DNA under torsion can be experimentally determined, and it usually increases with increasing contour length of the DNA and the magnitude of the pulling force. In the illustration shown in Figure 6-1b, the bead is rotated in a right-handed manner to form a positively supercoiled loop. The dependence of $<z>$ on *Lk* (in which $<z>$ is the average vertical projection of the extended DNA) can be readily measured at a particular *F*. Also, at a fixed *Lk*, a gradual increase in *F* would gradually increase $<z>$ and decrease the length of the supercoiled loop, progressively converting the writhe of the DNA within the original loop into twists in the DNA. These measured changes in turn provide information on the energetics of the twisting and writhing of a DNA. Furthermore, because at a given *F* the change in *Lk* of a DNA molecule held in a magnetic trap can be readily calibrated as a function of $<z>$, a change in the *Lk* by a DNA topoisomerase can be monitored, in real time, by observing the corresponding change in $<z>$.

---

5. See, for example, Strick, T.R., et al. 2000. *Nature* **404:** 901–904; the vertical force *F* can be calculated from $F = k_B T <z>/<\Delta x^2>$, in which $<\Delta x^2>$ is the mean square transverse fluctuation in bead position, $<z>$ is the average projection of the extended DNA in the vertical direction, *T* is the absolute temperature, and k_B is the Boltzmann constant equaling 1.3807×10^{-23} N m K^{-1}.

OPTICAL TRAP[6]

The use of one beam or two beams of laser light to trap and manipulate small particles was first shown in 1970, although the application of an optical trap (or optical tweezers) in the study of biological materials such as viruses and bacteria was not reported until much later.[6] The force exerted by a beam of light on a small particle can be understood in terms of the particulate nature of light, that is, a light ray consists of many photons moving at the speed of light in the direction of light propagation. When a ray "a" passes through a transparent bead with a refractive index higher than that of the medium the bead is in, the ray is bent toward the interior of the bead. A change in the direction of a light ray means a momentum change of all photons in it, and the law of momentum conservation requires a corresponding momentum change of the bead to offset the combined momentum change of all photos in the ray. This corresponding momentum change means a net force F_a on the bead, in the direction of the momentum change (Fig. 6-2).

The light ray striking a bead in a laser beam can be considered as a bundle of many pairs of parallel rays, each of which being symmetrical about the center of the bead, such as the pair *a* and *b* shown in Figure 6-2. Even though the rays *a* and *b* are symmetrical about the center of the bead, the forces F_a and F_b exerted by them are generally not equal in magnitude, because the intensity of a typical laser beam is the highest along its center, and decreases symmetrically toward the edges of the beam. Thus if a bead is located off the center of a light beam and ray *a* is closer to the beam axis than ray *b*, then the higher intensity of ray *a* relative to ray *b* means that $F_a > F_b$, and the *a* and *b* pair of light rays contributes a net force with two components: F_{grad}, the gradient component, points in the direction of increasing light intensity; F_{scat}, the scattering component, points in the direction of light propagation (Fig. 6-2). It is the component F_{grad} that tends to pull the bead toward the center axis of the light beam where light intensity is the highest; only if the bead is located exactly at the beam

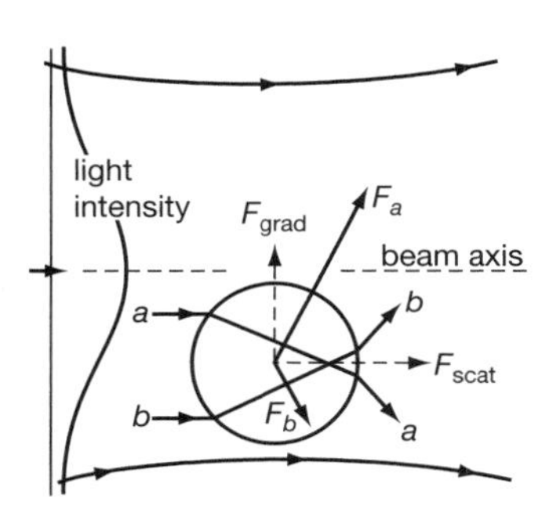

Figure 6-2. Forces experienced by a transparent bead in a light beam. The light beam propagates in the horizontal direction from left to right, and its intensity is the highest at the beam axis (dotted line) and decreases symmetrically toward the edges, as represented by the curve labeled "light intensity." The other symbols are defined in the text. (Redrawn, with permission, from Ashkin, A. 1997. *Proc. Natl. Acad. Sci.* **94:** 4853–4860, ©National Academy of Sciences U.S.A.)

6. Reviewed in Ashkin, A. 1997. *Proc. Natl. Acad. Sci.* **94:** 4853–4860.

center is $F_a = F_b$. The same argument goes for all the other pairs of rays that are symmetrical about the bead center, and the combined F_{grad} of all pairs of rays is responsible for pulling any off-axis bead toward the beam center.

The combined F_{scat} of all pairs of light rays would push the bead in the direction of light propagation in the absence of a light intensity gradient in that direction. Near the focal plane of a light beam, however, there is a steep gradient in light intensity along the direction of light propagation; hence this focused light beam can serve as an optical trap to hold the particle along both the light axis and the transverse directions. Alternatively, two opposing laser beams can be used to cancel the scattering components of the two light beams, and thus trap a transparent bead at an equilibrium position in their middle.

If a mechanical force F is applied to a bead held in an optical trap, through a DNA tethered to it for example, then the bead will be displaced from its equilibrium position in the absence of that force. Like the case of a spring, within a certain range of F the magnitude of the displacement d is proportional to the force, and the proportionality constant κ ($\kappa = d/F$) is termed the stiffness of the particular optical trap. Several methods have been developed to measure κ for a given optical trap; these include calibrating the stiffness of the trap by measuring d when a bead is subject to a known hydrodynamic drag force, calculating from the measured thermal Brownian motions of a trapped bead, calibrating through the observation of a known force-dependent structural transition in a macromolecule attached to the bead, or calculating directly from the changes in momentum fluxes of the laser light entering and exiting the trap. Once the stiffness of a particular optical trap has been calibrated, the known proportionality constant can in turn be used to calculate F from the experimentally observed displacement d.

For studying the actions of DNA topoisomerases on DNA, one end of the DNA is attached to a transparent bead held in an optical trap, and the other end is often attached to a bead mounted on the shaft of a computer-operated motor (Fig. 6-3). For the bead held in the optical trap, rotation around the vertical axis is countered by the hydrodynamic drag on the bead, and the DNA tethered to the bead can be positively or negatively supercoiled by rotating the DNA end anchored to the motor shaft.

HOP, STEP, AND JUMP

The mechanisms of different subfamilies of DNA topoisomerases, which we discussed in Chapters 3–5, predict a characteristic change in ΔLk for a single cycle of DNA breakage and rejoining by a type IA, IB, or II subfamily: The absolute value of ΔLk must change by 1 for a type IA enzyme, by 2 for a type II enzyme, and by an integer equal to or greater than 1 for a type IB enzyme. (We ignore the cases when Lk remains unchanged.) It would seem that these predicted changes in ΔLk per cycle of DNA breakage and rejoining should provide a straightforward test of the various mechanisms. However, owing

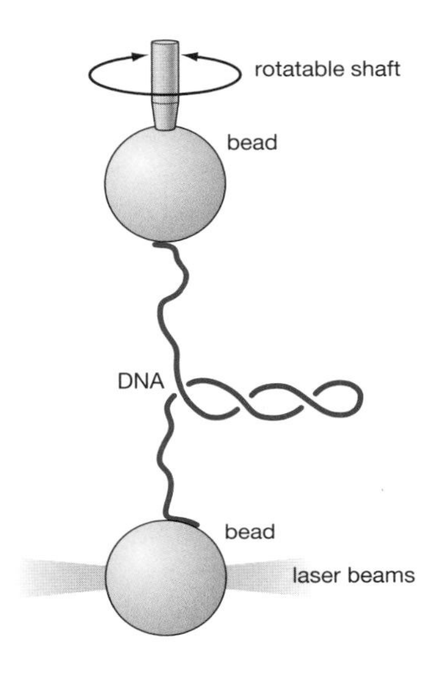

Figure 6-3. A bead-linked DNA in an optical trap. One end of the DNA is attached to a micron-sized transparent bead, and the other end is attached to a bead affixed to the shaft of a computer-controlled motor. A pair of optical tweezers is used to hold the transparent bead attached to DNA, and the computer-controlled motor is used to coil up the DNA. (Redrawn, with permission, from Charvin, G., et al. 2005. *Annu. Rev. Biophys. Biomol. Struct.* **34:** 201–219.)

to the lack of a net chemical change in the breakage and rejoining of DNA backbone bonds by a topoisomerase, a direct measurement of the number of cycles in a topoisomerase-mediated reaction is a formidable task. Only in the case of the type II DNA topoisomerases has the stepwise change of *Lk* been unequivocally shown: *Lk* has been shown to change in steps of 2 in the presence of ATP, and by only 2 when ATP is replaced by its nonhydrolyzable analog ADPNP, which presumably supports only a single cycle of reaction. For the type IA and IB DNA topoisomerases, earlier experiments showed that *Lk* changes by either odd or even integral numbers, but it had been difficult to conclusively determine the value of ΔLk per DNA breakage–rejoining cycle.[7]

The ability to observe changes in *Lk* in a single DNA molecule, in experiments using either a magnetic or an optical trap, can thus provide a real-time view of the action of a single DNA topoisomerase when the molar ratio of DNA topoisomerase to DNA is made sufficiently low to avoid multiple enzyme molecules acting on any one DNA molecule. Under such conditions, the dependence of *Lk* of a supercoiled DNA loop on time provides a powerful tool in inferring the action of a DNA topoisomerase.

Figure 6-4 illustrates an experiment with a type II DNA topoisomerase.[5] Here one digoxigenin-tagged end of an 11-kb-long DNA was fixed to a glass surface coated with antibodies against the plant steroid, and the other end of the DNA, tagged with biotin, was anchored to a streptavidin-coated bead held in a magnetic trap (depicted in Fig. 6-1).[4] A pulling force *F* measured in piconewtons (pN) is applied to the system, caus-

7. See, however, the paper by Stivers, J.T., et al. 1997. *Biochemistry* **36:** 5212–5222, in which evidence was presented that during each cycle of DNA breakage and rejoining the *Lk* of a supercoiled DNA was changed by multiple turns by vaccinia virus topoisomerase.

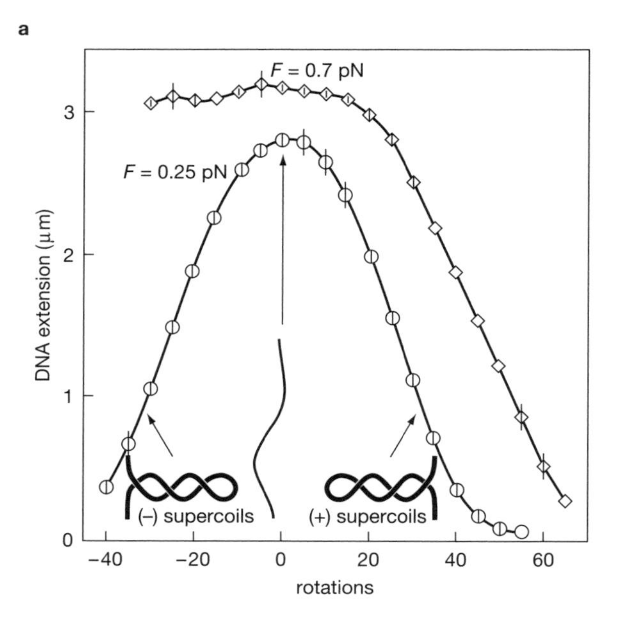

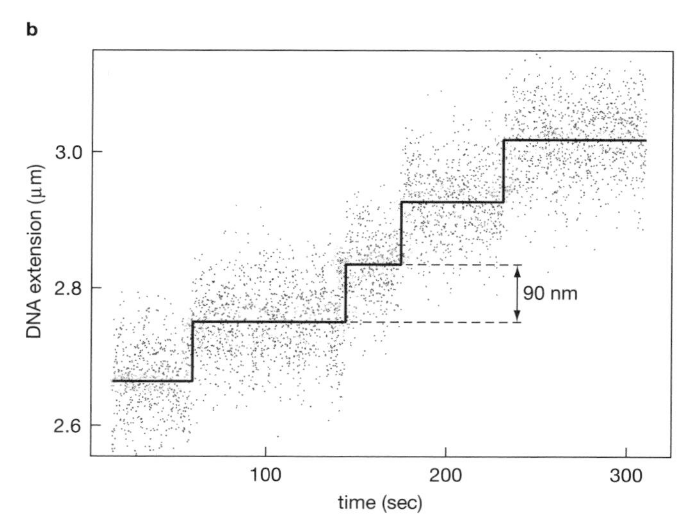

Figure 6-4. Removal of positive supercoils by a type IIA DNA topoisomerase. A magnetic trap (as depicted in Fig. 6-1) is used to manipulate an 11-kb DNA fragment. (*a*) The time-averaged *z*-extension of the DNA (<*z*>) is plotted as a function of the number of rotations of the pair of magnets. Rotating the DNA-linked paramagnetic bead in a right-handed way (positive numbers along the horizontal axis) leads to overwinding and positive supercoiling of the DNA, whereas rotating the bead in the opposite direction (negative numbers along the horizontal axis) leads to underwinding and negative supercoiling. (*b*) The vertical extension <*z*> of the DNA, overwound to form a positively supercoiled loop, is plotted as a function of time following the addition of a small amount of *Drosophila melanogaster* Top2. The dots represent raw data points, and the steps the best-fit average vertical extensions. See text for additional details. (Redrawn, with permission, from Charvin, G., et al. 2005. *Annu. Rev. Biophys. Biomol. Struct*. **34:** 201–219.)

ing rotation of the DNA-linked paramagnetic bead (a pN is one-trillionth of a newton, which is the force that increases the velocity of a 1-kg mass by 1 m/sec every second; the gravitational pull on a 1-kg mass is close to 10 N). This rotation led to positive or negative supercoiling of the DNA, as illustrated by the results of the calibrations depicted in Figure 6-4a. At a low pulling force $F = 0.25$ pN (lower curve), <z> is nearly symmetrically shortened by overwinding (right-handed rotations) or underwinding (left-handed rotations). At a higher pulling force of 0.7 pN (upper curve), however, the responses to overwinding and underwinding of the DNA were highly asymmetric. Whereas after several turns <z> was progressively reduced by overwinding; similarly to what was observed at the lower 0.25 pN, there was little change in <z> by underwinding at the higher F. The latter finding suggests that there is a change in the DNA helical structure on its negative supercoiling under a stretching force. In the experiment illustrated in Figure 6-4b, the removal of supercoils by *Drosophila melanogaster* DNA topoisomerase II (Top2) was examined under a pulling force of about 0.7 pN (the magnet was first rotated to form a positively supercoiled loop in the DNA before the addition of the enzyme). From data depicted in Figure 6-4a, we see that, for the particular experimental setup, <z> was linearly dependent on the number of rotations over a fairly broad range of positive supercoiling; within this range, an increment of 1 in the number of positive supercoils decreased <z> by 45 nm.

On the addition of a low amount of the topoisomerase (to avoid the presence of multiple enzyme molecules on a single positively supercoiled DNA) in the presence of ATP, stepwise increases in the vertical extension of an enzyme-bound DNA were observed over time (Fig. 6-4b). The observed step size, 90 nm, or twice the change in <z> per supercoil, strongly suggests that in each enzymatic cycle the type II DNA topoisomerase changes *Lk* by 2, as predicted from the results of numerous earlier "bulk" experiments with large populations of molecules. Significantly, steps corresponding to an increment of 45 nm in <z> were never observed. In this single-molecule experiment, the temperature was kept at 25°C and ATP was present at a low concentration of 10 μM to slow down the topoisomerase-catalyzed reaction. Under these conditions, the average time lapse between two successive stepwise changes in <z> was about 30 sec, which was sufficiently long relative to the time resolution of the apparatus used (~1 sec), and "bursts" showing *Lk* changes by multiples of 2 were rarely seen.

In the experiment illustrated in Figure 6-5, the action of a single DNA gyrase molecule was monitored by directly observing the axial rotation of a DNA through the use of a fluorescent "rotor" bead, strategically placed near a nick in the DNA, which serves as a swivel in the DNA (Fig. 6-5a).[8] The experimental setup (shown on the left) was otherwise very similar to that in Figure 6-3, with one end of the DNA affixed to a glass surface and the other anchored to a bead held in a magnetic trap. The action of a bound DNA

8. Gore, J., et al. 2006. *Nature* **439:** 100–104. For an earlier example of observing DNA axial rotation by the use of a fluorescent bead, see Harada, Y., et al. 2001. *Nature* **409:** 113–115.

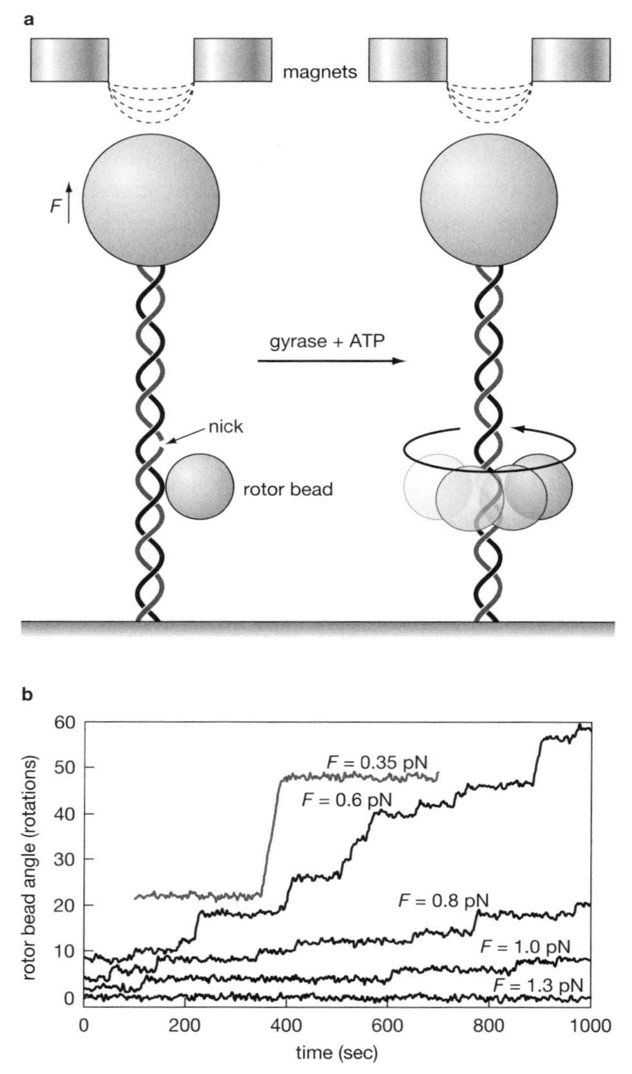

Figure 6-5. The use of a fluorescent bead to track the action of *E. coli* gyrase. The DNA fragment used here is 1.1 or 2.2 kb in length, with a strong gyrase binding site and a nick at a particular position. (*a*) Three pairs of molecular glues are used: one pair to attach one end of the DNA to a coated glass surface, a second pair to attach the other end of the DNA to a micron-sized magnetic bead, and the third pair to attach the DNA at the 3′-side of the nick to a 0.5-μm fluorescent bead. The fluorescent bead is used to monitor DNA rotation about its helical axis, catalyzed by action of gyrase in the presence of ATP (shown on the *right*). (*b*) The angular position of the rotor bead is shown as a function of time in the presence of *E. coli* gyrase and ATP. The rotor bead undergoes bursts of even-number rotations, the frequency and size of which are highly dependent on the pulling force *F*. For clarity, the tracings at different magnitude of *F* are arbitrarily displaced along the vertical axis. (Redrawn , with permission of MacMillan Publishers Ltd., from Gore, J., et al. 2006. *Nature* **439:** 100–104.)

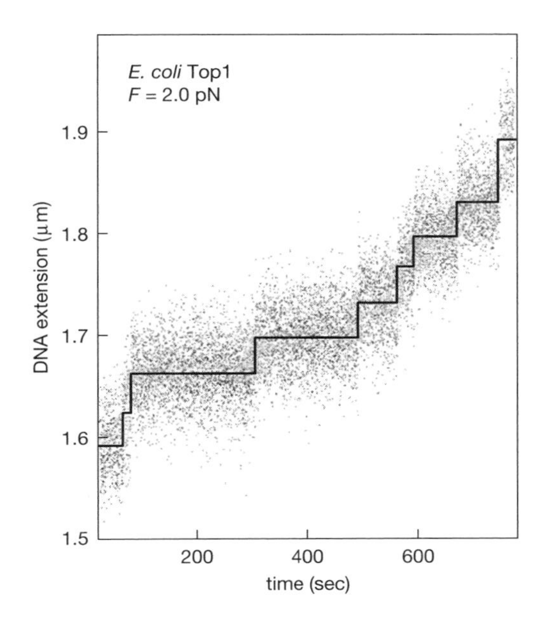

Figure 6-6. Supercoil removal by *E. coli* DNA topoisomerase I. To avoid the complication of helical structure changes when a negatively supercoiled DNA is stretched, a positively supercoiled loop was introduced into the 8.7-kb DNA held in the magnetic trap before the addition of the topoisomerase. The DNA used contained a 12-bp long segment in which the nucleotide sequences of the strands were not complementary to each other, which enhanced the removal of positive supercoils by a type IA DNA topoisomerase. The individual points are the raw data, and the lines the best-fit steps. (Redrawn, with permission, from Dekker, N.H. et al. 2002. *Proc. Natl. Acad. Sci.* **99:** 12126–12131, ©National Academy of Sciences U.S.A.)

gyrase can be monitored, in real time, by observing the rotation of the fluorescent bead (shown on the right). In the presence of *Escherichia coli* DNA gyrase and 1 mM ATP, the rotor bead was found to undergo bursts of rotations in multiples of 2 revolutions, as expected for the negative supercoiling of the DNA by gyrase that decreases *Lk* by 2 per reaction cycle. The burst size, however, turned out to be very sensitive to the pulling force. At an *F* of 1 pN, few bursts of more than 2 revolutions were seen; at an *F* of 0.8 pN or lower, burst sizes that are multiples of 2 revolutions were common (Fig. 6-5b). This strong dependence of burst size on *F* was interpreted in terms of a strong influence of the force on a kinetic competition between the wrapping of a DNA binding site around a gyrase, which commits the enzyme to a train of events, and the dissociation of the enzyme from the DNA, which terminates a train of events. In experiments at a higher resolution in time, what appeared as single bursts at a lower time resolution were each shown to consist of multiple steps of even-number changes in *Lk*. These single-molecule experiments thus provide a wealth of kinetic information not readily obtained in bulk experiments.

In contrast to the even number *Lk* changes accompanying reactions catalyzed by the type II DNA topoisomerases, single-molecule experiments with the type IA and IB enzymes displayed very different pictures. Figure 6-6 illustrates such an experiment with *E. coli* DNA topoisomerase I, a type IA enzyme, using a magnetic trap.[9] An 8.7-kb DNA containing a small single-stranded bubble, within which 12 unpaired nucleotides were present in each of the opposing strands, was used in this experiment. Under a stretch-

9. Dekker, N.H., et al. 2002. *Proc. Natl. Acad. Sci.* **99:** 12126–12131.

ing force of about 2 pN, the extended DNA was observed to progressively contract when the magnet was rotated right-handedly by more than about 20 revolutions, which signified the extrusion of a positively supercoiled loop of increasing length. (Overwinding an 8.7-kb DNA by 20 turns corresponds to positive supercoiling of a DNA to a specific linking difference of +20/[8700/10.5] or +0.024.) The magnitude of contraction per additional positive supercoil after the onset of the extrusion of the supercoiled loop was found to be 35.0 nm over a fairly broad range of positive supercoiling. Thus, within this range the removal of each positive supercoil by a type IA enzyme bound to the short single-stranded bubble would be expected to extend the DNA by 35.0 nm. The individual points in Figure 6-6 represent the observed changes in DNA extension at different times after the addition of the enzyme. Significantly, all experimental points can be fit by a line representing stepwise changes in DNA extension, with most of the individual steps corresponding to a single 35.0-nm increment, that is, the removal of a single supercoil in each step. Occasionally, events corresponding to the removal of two or more supercoils were observed. These events can be interpreted as successive events that occurred within the time resolution of the method (which is ~1 sec). A plot of the interval between the observed steps showed a Poisson distribution,[10] with an average interval of 15 sec between two successive events; for such a distribution, the chance that two successive steps occur within 1 sec is about 1 in 15, which is sufficiently high to account for the rare steps corresponding to *Lk* increments greater than 1.

When similar experiments were performed with the vaccinia virus enzyme, a type IB DNA topoisomerase, discrete stepwise changes in *Lk* were observed under conditions of a single enzyme acting on a DNA molecule. The distribution of ΔLk is well represented by an exponential, and the mean ΔLk calculated from this exponential distribution ranges from ~30 at a stretching force F of 0.2 pN to above 100 at $F = 3$ pN.[11,12] Thus in contrast to the case of a type IA DNA topoisomerase, the type IB vaccinia enzyme appears to permit multiple axial rotations of the DNA double helix during a single cycle of DNA breakage and rejoining by the enzyme.

BRAIDING A PAIR OF DNA DOUBLE HELICES

The experiments just described illustrate the power of a micromanipulator in the study of DNA topology and topoisomerases. Reactions such as the removal of DNA supercoils by a single topoisomerase molecule, or the ATP-dependent negative supercoil-

10. A Poisson distribution, first reported by Siméon-Denis Poisson (1781–1840), represents a distribution of the probability of a certain number (0, 1, 2, 3, ...) of discrete events occurring within a fixed time interval, when such events occur independently of each other with a particular average frequency.

11. Koster, D.A., et al. 2005. *Nature* **434:** 671–674.

12. In these measurements, because of the finite number of supercoils in a supercoiled loop, the maximal ΔLk of a step is constrained by this number; thus the measured distribution is inherently biased by this upper cutoff point unless corrected for this sampling error. See Koster, D.A., et al. 2006. *Proc. Natl. Acad. Sci.* **103:** 1750–1755.

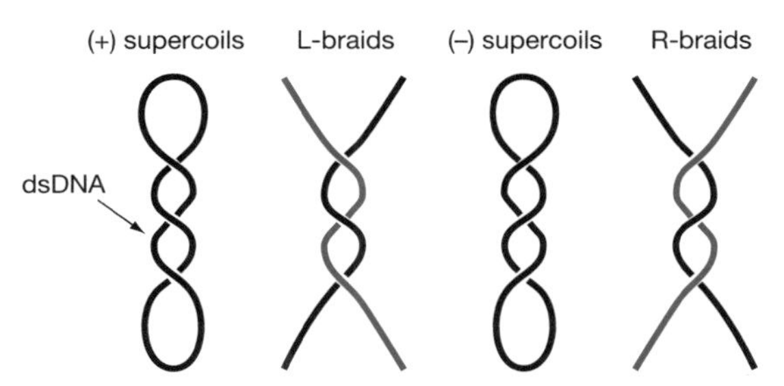

Figure 6-7. Structural similarity between the intertwined double helices in a positively supercoiled DNA and a pair of DNA double helices forming L-braids (*left* structures), and the same comparison for negatively supercoiled DNA and R-braids (*right* structures). (Redrawn, with permission, from Charvin, G., et al. 2003. *Proc. Natl. Acad. Sci.* **100:** 9820–9825, ©National Academy of Sciences U.S.A.)

ing of a DNA by a single gyrase, can be observed in real time in such experiments. Because the configuration of a DNA molecule can be deduced from the vertical projection of a DNA-attached bead confined in a magnetic or optical trap, these experimental approaches are particularly advantageous in studying some of the mechanistic aspects of DNA topoisomerase-catalyzed reactions. In the experiments on the stepwise removal of DNA supercoils by the type IA or IB DNA topoisomerase, for example, the difficulty of synchronizing a very large number of enzyme molecules would make the corresponding experiment in a bulk reaction rather challenging.

There are also other types of experiments that are difficult to perform in bulk reactions but can be performed readily through the use of micromanipulators. One example is experiments on the unbraiding of a pair of DNA double helices by a type II DNA topoisomerase.[13,14] In such an experiment, the two ends of a DNA are differently tagged with biotin and digoxigenin, and the digoxigenin-tagged ends are anchored to a glass surface coated with antibodies against digoxigenin. An avidin-coated paramagnetic bead is then attach to the biotin-tagged ends of a pair of DNA molecules, and a magnetic trap is used to rotate the bead one way or the other to intertwine the pair of the bead-attached DNA double helices into left-handed (L-) or right-handed (R-) braids.

A pair of braided linear DNA molecules, with their ends fixed to two surfaces, resembles a pair of intertwined double helices in a supercoiled DNA: The L-braids in a braided DNA resemble the plectonemic intertwines in a positively supercoiled DNA, and the R-braids those in a negative supercoil DNA (Fig. 6-7). The two cases differ, however, in that a single nick introduced into a supercoiled DNA loop would remove the supercoils, but nicking a braided pair of DNA double helices has no effect on the number of braids; whereas DNA supercoiling is tightly coupled to torsion in the double helix, in a braided pair of double helices this is not the case.

13. Stone, M.D., et al. 2003. *Proc. Natl. Acad. Sci.* **100:** 8654–8659.

14. Charvin, G., et al. 2003. *Proc. Natl. Acad. Sci.* **100:** 9820–9825.

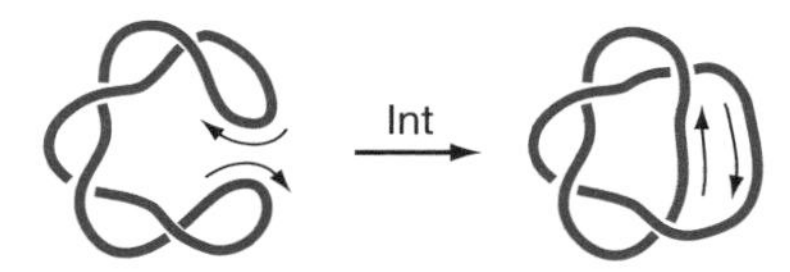

Figure 6-8. Use of site-specific recombination promoted by phage λ integrase (Int) to form a pair of R-braided rings. The lines represent double-stranded DNA and the arrows the recognition sites of the integrase. (Redrawn, with permission of Elsevier, from Boles, T.C. et al. 1990. *J. Mol. Biol.* **213:** 931–951.)

In Chapter 2, we considered the formation of pairs of multiply intertwined SV40 DNA rings, presumably owing to an inadequate level of the type II DNA topoisomerases during their duplication. The two DNA molecules in such a pair of rings are braided; furthermore, because the braids have their origin in the right-handed helical intertwines of the complementary strands in the parental DNA double helix, they are exclusively R-braids. Multiply intertwined rings with R-braids can also be formed by phage λ integrase promoted site-specific recombination between two sites on a negatively supercoiled DNA (Fig. 6-8).[15] Trapping the left-handed plectonemic intertwines in a positively supercoiled DNA ring into L-braids between the product rings cannot be performed in the same way, however, because the integrase-promoted reaction requires a negatively supercoiled DNA substrate (see Chapter 5).

Whereas only R-braided DNA rings can be readily formed and studied in bulk reactions, braids of either kind are readily introduced between a pair of DNA molecules attached to a single bead held in an optical or magnetic trap.[13,14] In the experiment depicted in Figure 6-9, a paramagnetic bead with a pair of attached double helices, each of which contained at least one nick, was rotated in a magnetic trap to braid the DNA molecules under a constant pulling force F of 1.15 pN.[14] The relative vertical extension of the bead, $<z>/<z_{max}>$ (in which $<z>$ is the z-extension and $<z_{max}>$ is the maximal extension at the particular pulling force) is plotted as a function of the number of revolutions (n) the bead has been rotated (negative and positive values of n denote counterclockwise and clockwise rotations, respectively, when viewed from the bottom). Braiding a pair of bead-attached DNA helices decreases $<z>/<z_{max}>$. As shown in Figure 6-9, the $<z>/<z_{max}>$ versus n plot is nearly symmetrical. There is a large drop in the relative extension for the first half turn of the bead, at which the two DNA molecules starts to cross, and a more gradual drop when more braids are formed (in the plot shown, the horizontal scale n is expended in the range $n = -1$ to $+1$). After about 70 turns have been introduced at the particular pulling force, the additional turns cause a steeper drop in the relative extension. This steeper drop of $<z>/<z_{max}>$ with increasing magnitude of n signifies further coiling of the tightly braided double helices in space, as depicted in the

15. Boles, T.C., et al. 1990. *J. Mol. Biol.* **213:** 931–951; Crisona, N.J., et al. 2000. *Genes Dev.* **14:** 2881–2892.

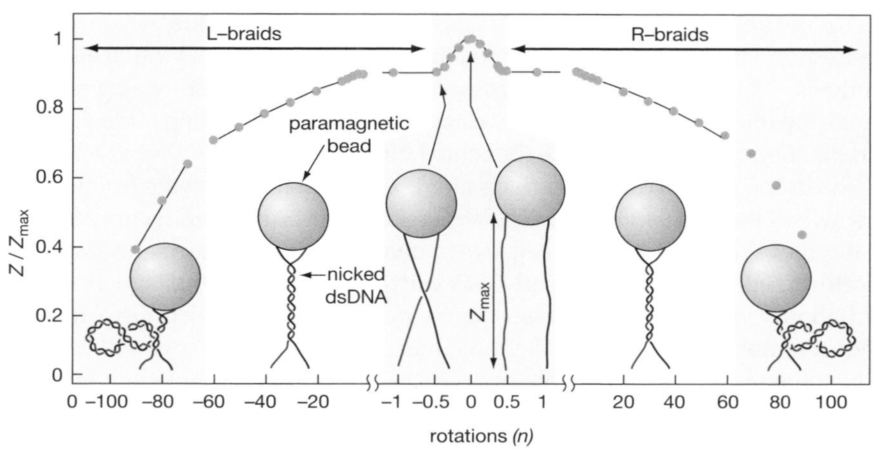

Figure 6-9. Braiding of a pair of DNA double helices in a magnetic trap. The data points are the experimentally measured $z(n)/z_{max}$, the DNA extension after n revolutions relative to the maximal extension of the pair of bead-attached DNA double helices; the drawings in the regions below the data points illustrate the corresponding conformations of the DNA pair. Note that the horizontal scale between $n = -1$ and $n = +1$ is expanded. See the text for details. (Redrawn, with permission, from Charvin, G., et al. 2003. *Proc. Natl. Acad. Sci.* **100:** 9820–9825, ©National Academy of Sciences.)

leftmost and rightmost panels of Figure 6-9. Increasing the number of L-braids between a pair of DNA double helices beyond a certain critical number leads to secondary coiling of the braided pair into a right-handed superhelix, whereas increasing the number of R-braids between a pair of DNA double helices beyond a certain critical number leads to secondary coiling of the braided pair into a left-handed superhelix. Onset of superhelix formation was found to occur when the vertical extension had contracted by about one third from $<z_{max}>$. The numerical value of this critical winding number is strongly dependent on the pulling force: The larger F is, the greater the winding number must be for the onset of superhelix formation of a braided DNA pair.

Once the dependence of the relative extension on n has been calibrated, unbraiding of the DNA by a type II DNA topoisomerase can be deduced by monitoring the time dependence of $<z>/<z_{max}>$ of a braided DNA pair at a particular initial value of n (n_0). With *Drosophila* Top2, at F of 1.15 pN and n_0 of either –40 or +40, the enzyme was found to unbraid the DNA at about the same rate. In either case, the rate of unbraiding was exponentially dependent on F, and extrapolates to about 2.9 braids per second at $F = 0$ under the experimental conditions used.

By contrast, *E. coli* DNA topoisomerase IV (Top4) appeared to unbraid L-braids much more efficiently than R-braids.[13,14] Unbraiding of L-braids proceeded rapidly until all braids are removed, but R-braids were essentially refractive to the action of this enzyme unless the number of R-braids exceeded a critical value, which was found to correspond to the onset of further coiling of the R-braided DNA into a left-handed superhelix.[14] Similar to the case of *Drosophila* Top2, the rate of untwining either L- or R-braids by the *E. coli* enzyme is also dependent on the stretching force *F*, but here the dependence is not a simple exponential function.

The strong preference of *E. coli* Top4 for L- over R-braids when the extent of braiding between the pair of DNA molecules is relatively low is reminiscent of its preferential removal of positive over negative supercoils: Single-molecule measurements indicate that positively supercoiled DNA with intertwines resembling L-braids is relaxed approximately 20 times faster than is negatively supercoiled DNA with intertwines resembling R-braids.[15] Earlier bulk measurements have shown, however, that the *E. coli* enzyme is also capable of removing R-braids to completion, even when both members of the braided rings are nicked.[16] These earlier measurements indicate that the rate of unbraiding a pair of nicked supercoiled DNA rings with R-braids is about an order of magnitude higher than the rate of negative supercoil removal, and is comparable to its rate in the removal of the positive supercoils.[16] Why do the results of the single-molecule and bulk experiments appear to differ in terms the removal of R-braids by the *E. coli* enzyme?

It is not straightforward to compare rates measured under very different conditions, because the rate-limiting steps in these measurements may not be the same.[17] The complex force dependence in the single-molecule unbraiding experiments with *E. coli* Top4 increases the difficulty of comparing the experimental results with those from bulk measurements. The conformational flexibility of the DNA segments, which may influence the rate of transporting one DNA double helix through another by a type II DNA topoisomerase, is also different for a pair of braided rings that are otherwise unconstrained and for a pair of braided DNA double helices being stretched by a force *F*.

Significantly, both bulk and single-molecule measurements indicate that supercoiling of a pair of braided DNA aids their unbraiding by *E. coli* Top4. In bulk experiments, the untwining of the R-braids between a pair of nicked DNA rings by the *E. coli* enzyme was found to increase by about a factor of 4 on negative supercoiling of the rings[16]; in single-molecule experiments, the *E. coli* enzyme is proficient in the removal of R-braids if the braided double helices are supercoiled.[18] The biological consequences of some of these properties of *E. coli* Top4 will be discussed in Chapters 7 and 8.

---

16. Ullsperger, C. and Cozzarelli, N.R. 1996. *J. Biol. Chem.* **271:** 31549–31556.

17. When a reaction proceeds sequentially through several steps, its overall rate is determined by the rate of the slowest (or rate-limiting) step. Altering the experimental conditions may affect the various steps differently; thus the same reaction performed under different experimental conditions may have a different rate-limiting step.

18. Reviewed in Postow, L., et al. 2001. *Proc. Natl. Acad. Sci.* **98:** 8219–8226. See also Kato, J., et al. 1990. *Cell* **63:** 393–404; Zechiedrich, E.L., et al. 1997. *Genes Dev.* **11:** 2580–2592.

WATCHING A LARGE MOLECULE DANCE

As we saw at the beginning of this chapter, the predicted large movements between different domains of a DNA topoisomerase, and between different parts of the enzyme and its bound DNA, make single-molecule FRET one of the most attractive methods for observing directly their actions. In the ATP-dependent transport of one DNA double helix through another by a type IIA DNA topoisomerase, for example, the opening and closing of the protein entrance and exit gate, as well as the opening and closing of the DNA gate, are likely to involve large displacements of the order of several nanometers. To monitor such long-distance movements in real time by FRET requires, however, the ability to place donor and acceptor fluorophores at appropriate locations in the enzyme and/or DNA.

Similar to tagging a DNA with biotin or digoxigenin, placing a fluorophore at a particular position in a long DNA or an oligodeoxyribonucleotide usually starts with the synthesis of a chemically modified cytosine (or thymine) with a chemically reactive group at its C5 position. Following incorporation of the modified nucleoside moieties into a DNA or oligonucleotide, an appropriate fluorophore can be coupled to the reactive group of the modified base. Because the two strands of an oligodeoxyribonucleotide can be separately tagged, one with a donor and other an acceptor fluorophore, annealing of the two differently tagged strands into a double helix provides a way of measuring, by FRET, the enzyme-mediated relative displacement between the D–A pair on the same DNA. If a D–A pair is placed on the two sides of DNA cleavage by a type II DNA topoisomerase, for example, then the enzyme-mediated opening and closing of the DNA gate can be monitored by the time dependence of FRET between the D–A pair.

Placing fluorophores at specific locations in a protein often uses the relatively reactive sulhydryl (SH) group in a cysteinyl residue. Site-specific mutagenesis can be used to substitute a cysteine for the amino acid residue normally present at a particular position of a polypeptide chain. It may be difficult to place a fluorophore at a particular cysteine within a protein possessing multiple cysteines, but even here a solution can usually be found. Yeast DNA topoisomerase II, for example, has a total of nine cysteines, but in fact none is essential, therefore all nine can be replaced by alanines without inactivating the enzyme.[19] To place a D–A pair on a single polypeptide, to study the relative movement between two domains, poses more of a chemical challenge. Sometimes, the two parts containing the intended tagging sites can be artificially separated (recall in Chapter 4, for example, the reconstitution of a functional human DNA topoisomerase I using one fragment containing the core domain region from residues 175 to 659, and the other the small carboxy-terminal domain spanning residues 713–765). If all else fails, a mixture of derivatives of D and A that have been prepared for coupling to cysteinyl residues can be used to randomly tag a pair of cysteines in the same

19. Lindsley, J.E. 1996. *Proc. Natl. Acad. Sci.* **93:** 2975–2980.

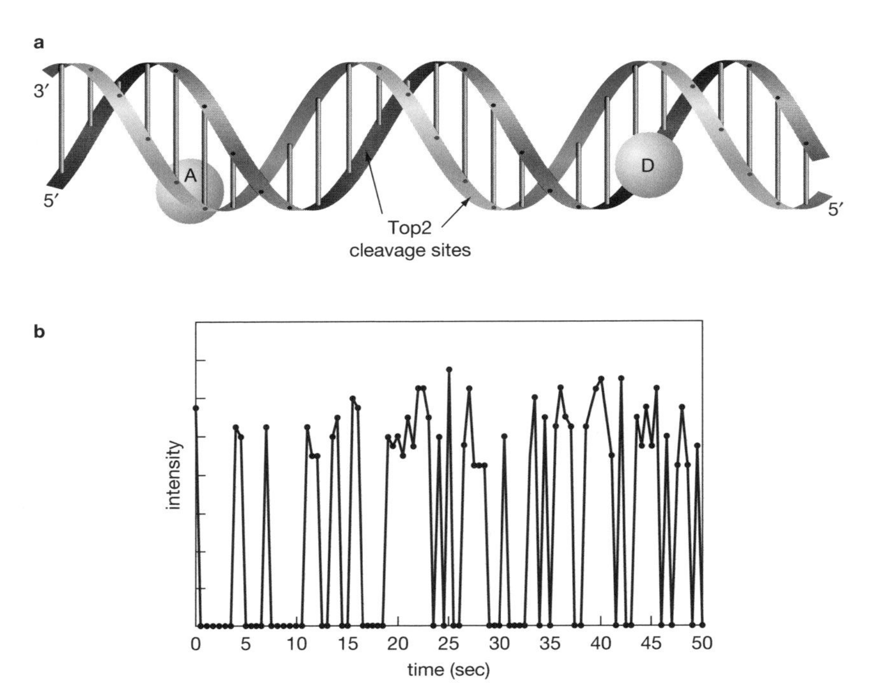

Figure 6-10. Monitoring the opening and closing of the DNA gate by *Drosophila melanogaster* DNA topoisomerase II with FRET. (*a*) The structure of the DNA used in this experiment is drawn according to the Watson–Crick representation of the DNA B-helical form; binding of the DNA may cause a significant bending of the DNA, however (for further details, see Dong, K.C. and Berger, J.M. 2007. *Nature* **450:** 1201–1205). The DNA contains a 28-bp region specifically bound by the *Drosophila* enzyme, with an extra A at one 5′-end and an extra G at the other 5′-end. The arrows indicate the locations of DNA cleavage, and the two spheres A and D represent, respectively, the locations of the acceptor and donor fluorophores. (*b*) FRET measurements are shown as a function of time in the presence of 50-nM *Drosophila melanogaster* DNA topoisomerase II, magnesium ions, and ATP. (Redrawn, with permission, from Smiley, R.D., et al. 2007. *Proc. Natl. Acad. Sci.* **104:** 4840–4845, ©National Academy of Sciences U.S.A.)

polypeptide. Individual molecules containing a D–A pair rather than two D or two A fluorophores can be selected during viewing by single-molecule FRET.

Figure 6-10a illustrates the features of a DNA fragment used in a single-molecule FRET experiment.[19] This DNA fragment contains a preferred *D. melanogaster* Top2 binding site, and the pair of arrows indicate the DNA breakage sites by the enzyme. The two spheres labeled D and A (the donor and acceptor fluorophore, respectively) were placed 15 bp or about 1.5 helical turns apart along the DNA. One of the DNA ends was tagged with a biotin, to facilitate attaching the DNA molecules to the surface of an avidin-coated glass slide. When *Drosophila* Top2 was bound to the DNA molecule in the presence of both ATP and magnesium ions, the observed fluorescence from FRET was found to fluctuate between a high-intensity state and a low-intensity state (Fig. 6-10b). But in the absence of the enzyme, ATP, or magnesium ions, however, no such fluctuations were seen, and the high intensity-state persisted.

The data shown in Figure 6-10b were interpreted in terms of the opening and closing of the DNA gate by *Drosophila* Top2: The high-intensity FRET state presumably corresponds to a closed gate with the D–A pair separated by about 51Å, and the low-intensity FRET state corresponds to an open gate state. From various fluorescence measurements, the distance between the two fluorophores in this open-gate state was calculated to be ~20 Å larger than that of the closed state.[19] From recordings for a total of about 200 individual molecules, the forward and reverse rate of interconversion between the high and low FRET state were estimated to be about 1 sec in the presence of the enzyme, ATP, and magnesium ions.[20] These results show the power of single-molecule FRET measurements in the elucidation of enzyme mechanisms.[21]

We have seen how the activity of a single topoisomerase molecule can be followed by using methods involving either micromanipulation or direct observation. These methods add a powerful tool in the study of the DNA topoisomerases, and they promise a wealth of mechanistic details heretofore unattainable. In the next chapter, we shall begin to consider the physiological roles of these extraordinary enzymes.

20. Smiley, R.D., et al. 2007. *Proc. Natl. Acad. Sci.* **104:** 4840–4845.

21. The finding that the type IIA DNA topoisomerase-mediated opening and closing of the DNA gate in a G-segment can occur in the absence of a T-segment is surprising. In Chapter 5, we suggested that the capture of the T-segment might be important in triggering the opening of the DNA gate. It is also surprising that the FRET results suggest that the rates of DNA gate opening and closing are about the same; as our earlier experiments suggested, the DNA gate remains mostly in the closed state. (See discussions in Wang, J.C. 2007. *Proc. Natl. Acad. Sci.* **104:** 4773–4774.)

CHAPTER 7

Manifestations of DNA Entanglement

Replication and Recombination

"DNA just is. And we dance to its music."

Richard Dawkins, *The Selfish Gene*, 1976, p. 133

THE WORKINGS OF THE DIFFERENT TYPES OF DNA topoisomerases have been described in the preceding several chapters. We have seen that all living organisms have at least one type IA DNA topoisomerase as well as one type II DNA topoisomerase of either the IIA or IIB category.[1] The type IB DNA topoisomerases, despite their conspicuous absence in some bacteria (notably in the celebrated *Escherichia coli*), are also widely present in a diverse spectrum of organisms. In both the budding and fission yeast, there exists one each of the three subfamilies IA, IB, and IIA. Two of each of these three subtypes are found in vertebrates, making a total of six. In the plant kingdom, there are even more of these enzymes, with all four subtypes IA, IB, IIA, and IIB well represented. Even some viral parasites come equipped with their own DNA topoisomerases, so that they need not rely solely on their hosts. The bacteriophage T4 encodes a type IIA enzyme, and the poxviruses encode a type IB enzyme. A large virus that infects amoebae, called the mimivirus, sets a viral record by encoding three DNA topoisomerases, one each of the subtypes IA, IB, and IIA.[2]

1. The archaeal phylum *Thaumarchaea* may constitute an exception to this generalization (Forterre, P., unpublished).

2. Raoult, D., et al. 2004. *Science* **306:** 1344–1350. The mimivirus isolated from the amoebae *Acanthamoeba polyphaga* has a chromosome 1.2 million bp in size.

Why does a living organism go to the trouble of producing multiple DNA topoisomerases? Do different subfamilies of these enzymes, or even different members of the same subtype, act in different cellular processes? Are all six human or mouse DNA topoisomerases indispensable, or are some perhaps functionally redundant, so that any one of these could also shoulder the responsibilities of its congener?

One of the most useful approaches in elucidating the physiological function of a cellular component is genetic analysis through inactivation or modification of the gene encoding this component. A collection of mutant mice, each mouse lacking one of the six mammalian DNA topoisomerases, was obtained in the mid-1990s and have been closely examined since.[3–8] These studies suggest that among the five mammalian enzymes that reside primarily inside the cell nucleus (DNA topoisomerases I, IIα, IIβ, IIIα, and IIIβ encoded, respectively, by the genes *TOP1*, *TOP2α*, *TOP2β*, *TOP3α*, and *TOP3β*) each one is indispensable and presumably has a unique physiological role. Mouse embryos lacking either a functional *TOP1* or *TOP2α* gene die very early, before a fertilized egg gets to the eight-cell stage.[3,7] Mouse embryos lacking a functional *TOP3α* gene survive a bit longer, but they, too, perish around the time of their implantation into the uterus wall (about 4 d following the fertilization of an egg).[4] Mouse embryos bearing a deletion in the *TOP2β* gene can develop to term, but they die shortly after birth owing to a breathing impairment caused by neural and neuromuscular defects.[5] Even though mice without a functional *TOP3β* gene can develop to maturity with no apparent defects, these $top3\beta^{-/-}$ mice show a progressive decrease in fecundity over time, as well as from one generation to the next.[6] (The term $top3\beta^{-/-}$ denotes the genotype of a mutant mouse, having both copies of its *TOP3β* gene inactivated.) The $top3\beta^{-/-}$ mice also develop autoimmunity as they age, and acquire lesions in multiple organs from inflammatory responses.[6] Probably as a consequence of this immunological defect, the average lifespan of the $top3\beta^{-/-}$ mice is only about one-half as long as their siblings with at least one intact copy of the *TOP3β* gene.[6] The only mouse *TOP* gene that appears to be dispensable with no significant physiological consequence is *TOP1mt* that encodes the mitochondrial type IB DNA topoisomerase.[8] As we shall see in Chapter 9, the metabolically active DNA molecules inside mitochondria are ring-shaped and replicate in the ring form. Duplication of mitochondrial DNA, therefore, is likely dependent on the presence of at least one DNA topoisomerase. This line of reasoning suggests that one or more of the

3. Morham, S.G., et al. 1996. *Mol. Cell Biol.* **16:** 6804–6809.
4. Li, W. and Wang, J.C. 1998. *Proc. Natl. Acad. Sci.* **95:** 1010–1013.
5. Yang, X., et al. 2000. *Science* **287:** 131–134; Lyu, Y.L. and Wang, J.C. 2003. *Proc. Natl. Acad. Sci.* **100:** 7123–7128.
6. Kwan, K.Y. and Wang, J.C. 2001. *Proc. Natl. Acad. Sci.* **98:** 5717–5721; Kwan, K.Y., et al. 2003. *Proc. Natl. Acad. Sci.* **100:** 2526–2531.
7. Akimitsu, N., et al. 2003. *Genes Cells* **8:** 393–402.
8. Reviewed in Zhang, H., et al. 2007. *Biochimie* **89:** 474–481.

nuclear mammalian DNA topoisomerases can enter the mitochondria and substitute for the mitochondrial enzyme.

These mouse gene disruption experiments illustrate the challenges in understanding the physiological roles of an enzyme that can potentially participate in multiple cellular processes. Why does the inactivation of an enzyme like DNA topoisomerase IIβ, which catalyzes the passage of one DNA double helix through another (see Chapter 5), affect mainly neural and neuromuscular development? Why does the absence of DNA topoisomerase IIIβ lead to the development of autoimmunity?

In addressing these questions, the mechanisms of the DNA topoisomerases and the known characteristics of various cellular transactions of DNA provide a general framework for thinking about the plausible biological roles of these enzymes. Additional information comes from the locations of the various DNA topoisomerases in cells, the proteins they interact with, how their cellular concentrations vary during the stages a cell passes through and during differentiation, etc. Once all clues from these different observations are combined and considered, the physiological consequences of inactivating a *TOP* gene encoding a particular DNA topoisomerase become less bewildering. Over the past several decades, a picture of how the various DNA topoisomerases function inside different living organisms has gradually emerged, and is likely to become increasingly more clear as more experimental data accumulate.

For simplicity and convenience, we shall consider the roles of the DNA topoisomerases in DNA replication, recombination and repair in separate sections in this chapter. And in Chapter 8, we shall discuss the importance of DNA topoisomerases in gene expression. Inside a cell, however, the various transactions of DNA are usually interrelated and often occur at the same time. A common thread connecting the roles of the DNA topoisomerases in all these processes is DNA entanglement, a problem deeply rooted in the structure of the DNA double helix.

THE ROLES OF DNA TOPOISOMERASES IN REPLICATION

Entanglement of the Parental DNA Strands during Replication

As the entanglement problem accompanying DNA replication is historically our focus, we consider this process first, including a few topological issues that were omitted in our earlier discussions. Recall that, during replication, the base-paired DNA strands must become uncoiled to separate. Figure 7-1 illustrates a DNA molecule undergoing replication with one end fixed in space, and, at the other end, its two complementary strands separated at the point called the replication fork. But as the intertwined strands separate, the region ahead of them becomes more tightly wound—the helical turns that were between the strands before their parting are now

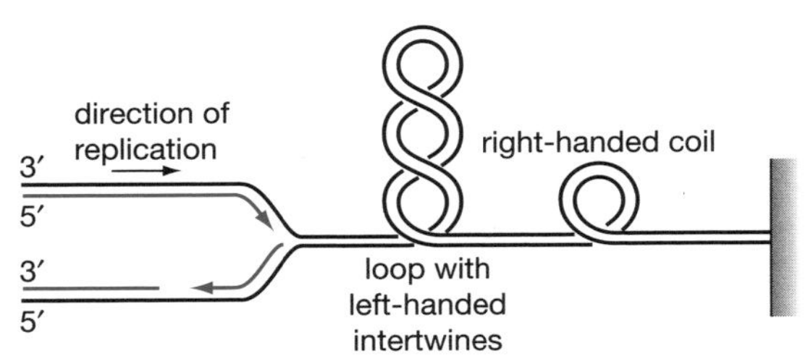

Figure 7-1. Positive supercoiling of a DNA segment ahead of a replication fork. For simplicity, the plectonemic DNA double helix is represented by two parallel, rather than twisted, lines. The black lines here represent the parental DNA strands, and the blue ones the newly synthesized progeny strands. The arrowheads show the direction of elongation of newly synthesized strands. For a general discussion on DNA replication, consult an introductory biochemistry or molecular biology textbook (see, for example, Chapter 7 in Watson et al. 2007. *Molecular biology of the gene*, 6th ed. Cold Spring Harbor Laboratory Press, Cold Spring Harbor, New York).

squeezed into the shortened region ahead. This overwinding leads to positive supercoiling of the DNA, and, if there is sufficient slack in the DNA ahead of the fork, the over-twisted molecule would writhe in space along a *right-handed* trajectory, or the DNA may double on itself to form a hairpin loop with *left-handed* intertwines (Fig. 7-1; see also Fig. 2-3 in Chapter 2).[9]

As we saw in Chapter 1, the replication of a DNA molecule requires its base-paired strands to uncoil and separate. The region where the parental strands come apart and the progeny strands are extended, called the replication fork, is shown advancing from left to right in Figure 7-1, and the arrowhead on each newly synthesized progeny strand indicates the direction of its elongation. Because the DNA replication machinery (a "replisome") always adds nucleotides to the 3′ end of an elongating progeny strand, copying of the upper parental strand shown in the figure can proceed continuously in the same direction as fork movement, but copying of the other strand would move the growing 3′ end farther from the point where the parental strands come apart; thus synthesis of the other strand must periodically restart from an upstream location. These two progeny strands are termed the leading and lagging strand, respectively. With continual elongation of the progeny strands, the replication fork progresses along the DNA, forcing overwinding and positive supercoiling of the DNA ahead of the fork.

9. That DNA overwinding can lead to contortions of opposite handedness may seem odd (the same is true with DNA underwinding, which would lead to the formation of loops with right-handed intertwines or left-handed coils). It should be noted, however, that the handedness of a spatial curve refers to a particular axis. For two DNA segments going up and down the hairpin loop shown in Fig. 7-1, they are intertwined left-handedly about a vertical axis in the plane of the paper that bisects the loop; in the right-handed coil shown in the same figure, the reference axis is one perpendicular to the plane of the paper and passes through the center of the coil.

In quantitative terms, unzipping every 10 base pairs of the complementary strands would introduce one positive supercoil in the base-paired region ahead of the fork. During replication, if the positive supercoils were allowed to accumulate ahead of a replication fork, further separation of the DNA strands would become progressively more difficult, and fork progression and further lengthening of the newly synthesized DNA strands would be severely retarded.

In principle, positive supercoiling of the DNA ahead of the parting strands can be avoided in two very different ways: either by allowing the anchored end depicted in Figure 7-1 to twirl in space, or by resorting to an enzymatic process capable of removing the supercoils. Twirling one end of a replicating DNA is feasible only if the DNA is relatively short; otherwise, the action of one or more DNA topoisomerases, especially a type IB or type II enzyme that is proficient in the removal of positive supercoils, would be necessary.

Experiments with budding yeast strains bearing mutated DNA topoisomerase genes show that Nature has indeed opted for an enzymatic solution of the DNA uncoiling problem in replication.[10] As we have seen, yeast has a total of three DNA topoisomerases, one and only one in each of the type IA, IB, and IIA categories. The genes *TOP1*, *TOP2*, and *TOP3* encode, respectively, the type IB, IIA, and IA enzymes I, II, and III (recall we refer to these enzymes as Top1, Top2, and Top3). Yeast strains carrying mutations (either singly or in combination) in the genes *TOP1*, *TOP2*, and *TOP3* have been used to show that, in the absence of both Top1 and Top2, elongation of newly synthesized DNA chains is blocked when the chain length gets to about 10,000 nucleotides. But, if either Top1 or Top2 is active, then chain elongation is nearly normal.[11] Thus in yeast either the type IB or type IIA DNA topoisomerase can fulfill the role of removing the positive supercoils during replication, but the type IA enzyme (Top3) is incapable of supporting the elongation of the nascent DNA chains.

In systems other than the budding yeast, the type IA enzymes are also generally thought to be ineffective in facilitating the untwining of the parental DNA strands. There appear to be, however, two notable exceptions. First, an in vitro system composed of purified proteins for the replication of an *E. coli* plasmid pBR322 has been shown to fully support the replication of this plasmid to form unlinked progeny rings. And, in this system, replication occurs even when the type IA enzyme *E. coli* Top3 is the only DNA topoisomerase present in the system.[12] Second, decreasing the cellular level of a mitochondrial type IA DNA topoisomerase of the unicellular protozoan *Trypanosoma brucei* has been observed to interfere with the late stage replication of a

10. See, for example, Uemura, T. and Yanagita, M. 1984. *EMBO J.* **3:** 1737–1744; Brill, S.J., et al. 1987. *Nature* **326:** 414–416. Earlier studies supporting this view were reviewed in Gellert, M. 1981. *Annu. Rev. Biochem.* **50:** 879–910; Wang, J.C. 1985. *Annu. Rev. Biochem.* **54:** 665–697.

11. Kim, R.A. and Wang, J.C. 1989. *J. Mol. Biol.* **208:** 257–267.

12. Hiasa, H. and Marians, K.J. 1994. *J. Biol. Chem.* **269:** 32655–32659.

class of small mitochondrial DNA rings.[13] Whereas the inability of a type IA enzyme to remove positive supercoils makes it an unlikely participant in the untwining of the parental DNA strands, the presence of single-stranded regions at a replication fork might change this situation. As we discussed in Chapter 3, a type IA enzyme is fully capable of relaxing a positive supercoiled DNA if there is a single-stranded region in it.

In organisms possessing multiple type II DNA topoisomerases, it appears that one of them often has a predominant role in the removal of the positive supercoils generated by advancing replication forks. In *E. coli*, gyrase apparently shoulders this crucial function, whereas DNA topoisomerase IV (Top4) may augment gyrase in this role. The best evidence for this division of labor comes from a study of the rates of replication fork progression in various thermal sensitive mutants of *Salmonella typhimurium*.[14] It was shown that thermal inactivation of a gene encoding the A subunit of Top4 had very little effect on fork movement. By contrast, heat inactivation of the gene encoding the A subunit of gyrase led to a slow cessation of fork progression, and heat inactivation of the A subunits of both gyrase and Top4 led to rapid cessation of fork progression.[14] That gyrase, rather than Top4, is the primary enzyme in positive supercoil removal is also consistent with the known catalytic properties of these enzymes. Gyrase has a robust activity in positive supercoil removal, whereas Top4, by comparison, is much less proficient in this reaction (see our earlier discussions in Chapters 5 and 6). In mammalian cells, two lines of evidence suggest that, besides Top1, the type IIA enzyme Top2α plays a more significant role than does Top2β in positive supercoil removal during DNA replication. First, Top2α and not Top2β is preferentially expressed in proliferating cells.[15] Second, purified human Top2α removes positive supercoils at a much faster rate relative to its removal of negative supercoils, whereas the Top2β isozyme shows no such specificity.[16]

An interesting possibility, first pointed out in 1980 by J. J. Champoux and M. Been, is that positive supercoils in the unreplicated DNA ahead of a replication fork might be converted to intertwines or "braids" (see Chapter 6) between the pair of newly replicated DNA segments behind the fork (Fig. 7-2a), and these braids could then be removed by passing one DNA double helix of the braided pair through the other.[17–19] Whether this conversion could actually occur is most likely to depend on how the repli-

13. Scocca, J.R. and Shapiro, T.A. 2008. *Mol. Microbiol.* **67:** 820–829.

14. Khodursky, A.B., et al. 2000. *Proc. Natl. Acad. Sci.* **97:** 9419–425.

15. Nitiss, J.L. 1998. *Biochim. Biophys. Acta.* **1400:** 63–81.

16. McClendon, A.K., et al. 2005. *J. Biol. Chem.* **280:** 39337–39345.

17. Champoux, J.J. and Been, M.D. 1980. Topoisomerases and the swivel problem. In *Mechanistic studies of DNA replication and genetic recombination: ICN-UCLA Symposia on Molecular and Cellular Biology* (ed. B. Alberts), Vol. 19, pp. 809–815. Academic Press, New York.

18. For reviews, see Postow, L., et al. 2001. *Proc. Natl. Acad. Sci.* **98:** 8219–8226; Espeli, O. and Marians, K.J. 2004. *Mol. Microbiol.* **52:** 925–931.

19. The term "braid" is used here for a twisted pair of DNA double helices; in everyday language, braiding generally refers to interweaving three or more strings or bundles of strings.

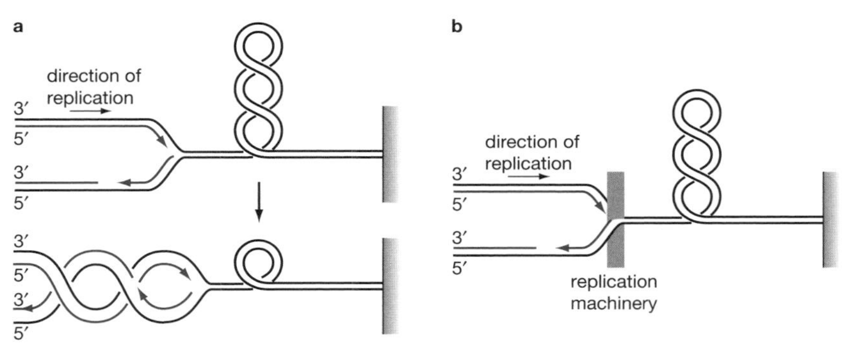

Figure 7-2. Conversion of positive supercoils ahead of a replication fork. (*a*) Here positive supercoils are converted to right-handed intertwines (R-braids) between the newly replicated DNA duplexes. (Redrawn, with permission, from Postow, L., et al. 2001. *Proc. Natl. Acad. Sci.* **98:** 8219–8226, ©National Academy of Sciences U.S.A.) (*b*) In this scheme conversion of positive supercoils requires rotation of the replication apparatus at the fork, here represented by a rod passing in between the pair of progeny duplexes, around the helical axis of the downstream parental DNA segment.

cation machinery is organized at a replication fork. In Figure 7-2b, the replication machinery for elongating the new DNA is represented by a short rod. Clearly, converting the positive supercoils shown in Figure 7-1 to the duplex braids shown in Figure 7-2a requires that the rod revolve about the helical axis of the unreplicated DNA ahead of the fork. Several studies have suggested, however, that the replication machineries may be anchored in the nuclear membrane of a eukaryotic cell or in the cell membrane of a prokaryotic organism.[20] If so, then the conversion of the positive supercoils in the parental DNA segment to duplex intertwines between the progeny DNA segments would be severely hindered.

It also appears that the removal of positive supercoils ahead of the replication fork is sufficient for fork progression, and it does not seem necessary to first convert the positive supercoils ahead of the fork to braids behind the fork. As we saw above, the type IB enzyme Top1 in the budding yeast can support DNA elongation in the absence of Top2. If conversion of the positive supercoils to braids between two progeny DNA segments were necessary for their removal, then the type IB enzyme (which can efficiently remove positive supercoils but cannot unbraid a pair of intertwined duplex DNA segments) would not be capable of supporting nascent DNA chain elongation.

It is probably also significant that the conversion of positive supercoils to duplex intertwines produces right-handed or R-braids rather than left-handed or L-braids (see Fig. 7-2a). These R-braids are likely to resemble more closely those between a pair of nicked, rather than supercoiled, DNA double helices, owing to the presence of short single-stranded regions in the newly duplicated DNA.[19] Whereas eukaryotic type IIA

20. Hozak, P. and Cook, P.R. 1994. *Trends Cell Biol.* **4:** 48–52.

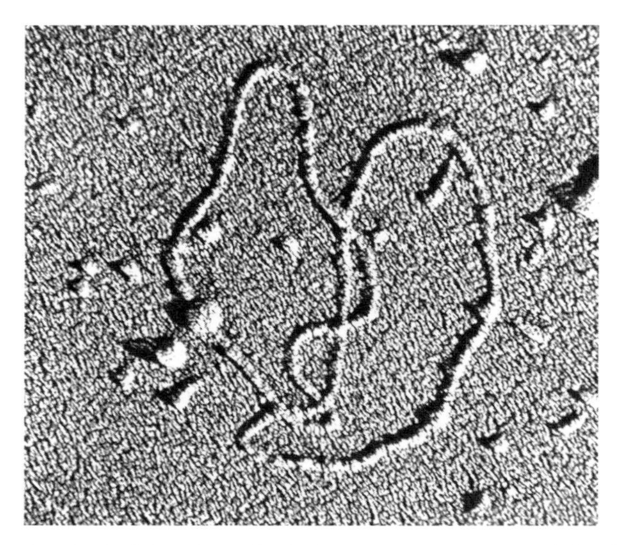

Figure 7-3. A pair of multiply linked SV40 DNA rings. (Electron micrograph kindly provided by A. Varshavsky.)

DNA topoisomerases such as *Drosophila* Top2 removes L- and R-braids with equal efficiency, in *E. coli* the main unbraiding activity Top4 appears to be less efficient in unbraiding R-braids between a pair of nicked DNA double helices (see Chapter 6).

Delbrück's Prophecy Came True

Recall from Chapter 1 that Max Delbrück's objection to the double helix model of DNA, posed in 1953, was based on the notion that replication of such a structure would form a pair of braided double helices. Nearly three decades later, his prophecy became a reality. In 1980, in the laboratory of Alexander Varshavsky, monkey cells infected with a simian virus called SV40 were accidentally cultured in a medium overly rich in nutrients and salts. The medium therefore became "hypertonic," that is, the total molar concentration of all membrane-permeable molecules in the medium became higher than that inside a normal cell. When the ring-shaped SV40 DNA was isolated from cells cultured in this hypertonic medium, the normally monomeric DNA was found to be in the form of a peculiar dimeric catenane in which a pair of monomeric rings are wrapped around each other, forming up to some 20 braids (Fig. 7-3).[21]

It is quite common for a ring-shaped DNA, such as SV40 DNA or the much larger *E. coli* chromosome, to start replication from a unique region in the DNA termed the

21. Sundin, O. and Varshavsky, A. 1980. *Cell* **21:** 103–114; 1981. **25:** 659–669.

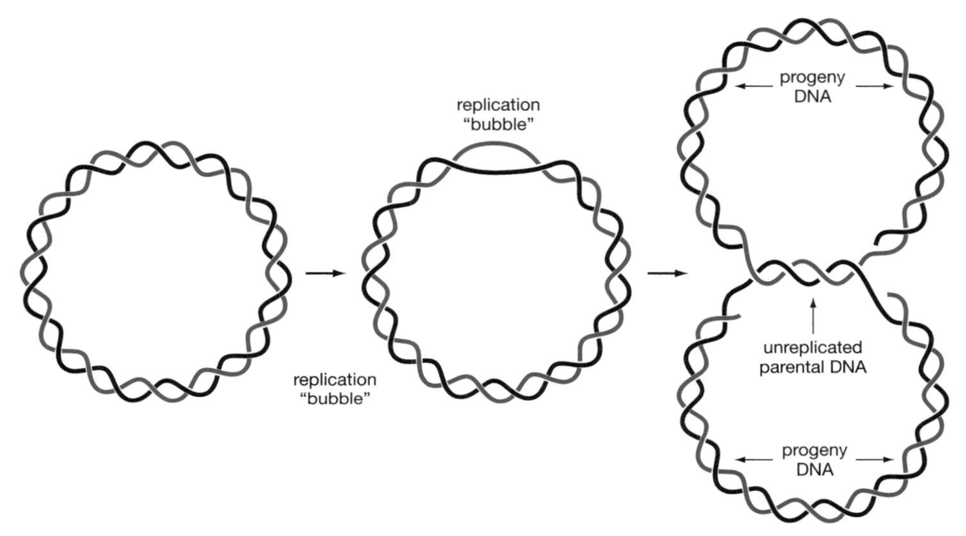

Figure 7-4. Replication of a DNA ring from a unique origin of replication. The replication machinery first opens a short DNA segment within a particular region of the DNA to form a replication "bubble" (*middle* drawing); copying of the parental strands initiates within this bubble. Replication progresses, often with a pair of replication forks moving bidirectionally.

replication origin. A small "bubble" is first formed at this location, on initiation of replication and separation of a short segment of the parental strands (Fig. 7-4). This replication bubble gradually increases in size as DNA synthesis proceeds, often in both directions along the DNA ring, and replication ends when the two replication forks converge near the point antipodal to the replication origin. Varshavsky and O. H. Sundin suggested that, when the length of the unreplicated DNA double helix between a pair of converging forks along an SV40 DNA ring is reduced to about 200 bp, further fork progression might convert these helical turns in the parental DNA into intertwines or braids between the duplicated progeny DNA molecules, as Delbrück first envisioned (Fig. 7-5; see also Fig. 1-3 in Chapter 1). Under normal conditions all intertwines would be removed by a type II DNA topoisomerase to yield a pair of separate monomer-sized rings. But, in the hypertonic medium the enzyme is presumably much less active, which in turn leads to the formation of multiply catenated SV40 DNA dimers.[21] In support of the idea that the catenation of SV40 DNA in a hypertonic medium is caused by a reduction in cellular type II DNA topoisomerase activity, plasmid DNA rings in mutant yeast cells expressing a thermal sensitive (ts) DNA topoisomerase II were found to form multiply intertwined catenated dimers (braided pairs of rings) when the growth temperature was up-shifted to inactivate the enzyme.[22]

22. DiNardo, S., et al. 1984. *Proc. Natl. Acad. Sci.* **81:** 2616–2620.

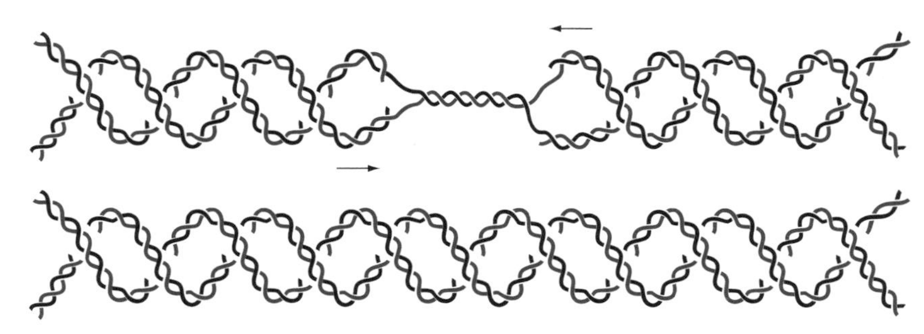

Figure 7-5. Conversion of the right-handed parental intertwines to R-braids between the progeny DNA molecules at the final stage of replication. For further details, see Sundin, O. and Varshavsky, A. 1980. *Cell* **21**:103–114; 1981. **25:** 659–669.

Type II DNA Topoisomerase and Chromosome Segregation

A braided pair of DNA rings fails to come apart unless a type II DNA topoisomerase unlinks them. But what about a linear chromosome? Does chromosome segregation in general require a type II DNA topoisomerase?

There is strong evidence, mainly from the use of various topoisomerase or *top* mutants of the fission yeast *Schizosaccharomyces pombe* and the budding yeast *Saccharomyces cerevisiae*, that the requirement of a type II DNA topoisomerase in chromosome segregation is not limited to ring-shaped chromosomes. In one example,[23] a double mutant of *S. pombe* (genotype *top2 ts nda3 cs*) was used. This mutant expresses a thermal sensitive (ts) Top2 (encoded by *top2 ts*) that is active at 20°C but inactive at 36°C, and a cold sensitive (cs) β-tubulin (encoded by a mutated *nda3* gene,

For eukaryotic organisms in particular, each round of cell division passes through a cycle of distinct stages in an orderly and intricately regulated way, termed the cell cycle. DNA replication occurs in a period of the cell cycle called the S phase (S for synthesis), during which replication starts at many origins in each chromosome, and adjacent replication forks eventually merge after traversing the intervening DNA. Segregation of the duplicated chromosomes and cell division occur during a period termed the M phase (M for mitosis, also known as "interphase"). S and M are interspersed by the gaps G_1 and G_2, in which G_2 is the period between S and M, and G_1 the period between M of one cell cycle and the S of the next cell cycle. The newly duplicated chromosome pairs (called sister chromatids) stay together and become highly condensed during the early part of M. They are then aligned at the mitotic plate in the middle of a dividing cell, and pulled to opposite cell poles by microtubules extending from the mitotic spindles at the poles.

23. Uemura, T., et al. 1987. *Cell* **50:** 917–925.

nda3 cs) that is inactive at 20°C but active at 36°C. A culture of the double mutant was first kept for several hours at 20°C, a temperature at which the cs β-tubulin was inactive and consequently the formation of mitotic spindle was blocked. Under these conditions, chromosome duplication and condensation were apparently unaffected by the lack of a functional mitotic spindle, and rodlike chromatids were visible under a light microscope. Segregation of the duplicated and condensed sister chromatids could not proceed at this temperature, however, because no mitotic spindle was present to pull them apart. On shifting the temperature to 36°C to reactivate the mutant β-tubulin, spindle function was restored, and the cells were found to attempt chromosome segregation. At 36°C, however, Top2 was inactive in the mutant cells. At the higher temperature, chromosome segregation during M was observed to be aberrant in the absence of active Top2, and a significant fraction of the mutant cells showed connected chromosomes that were being stretched and streaked along the mitotic spindle. From these observations we can deduce that, even at 20°C (and thus in the presence of a functional Top2 in the mutant cells), the fully replicated sister chromatids remain entangled to some extent. That they remain linked in the presence of an adequate level of Top2 is probably the consequence of random entanglement between uncondensed chromosomes in the confined space of a nucleus. These linked chromosomes cannot be readily pulled apart by the mitotic spindles in the absence of Top2, despite their linear (rather than ring) form. Thus the continued presence of an active Top2 during chromosome segregation is necessary for proper partitioning of the duplicated chromosomes.

Other studies in various organisms ranging from bacteria to mammals, using specific mutants and/or inhibitors of DNA topoisomerases and/or microtubule activities, have provided additional strong support to the idea that segregation of the entangled chromosome pairs requires the action of at least one type II DNA topoisomerase.[24,25] Especially in the case of long chromosomes, removing all intertwines between a pair of newly replicated chromosomes by twirling the chromosomal ends is apparently not sufficiently robust to ensure the complete separation of the sister chromosomes. Thus a high frequency of chromosome loss or chromosome breakage may occur when a tangled chromosome pair is pulled apart.[26]

24. See, for example, Uemura, T. and Yanagida, M. 1984. *EMBO J.* **3**: 1737–1744; Holm, C., et al. 1985. *Cell* **41:** 553–563; Uemura, T. and Yanagita, M. 1986. *EMBO J.* **5:** 1003–1010; Kato, J., et al. 1990. *Cell* **63:** 393–404; Ishida, R., et al. 1994. *J. Cell Biol.* **126:** 1341–1351.

25. To facilitate the precise partitioning of two sets of chromosomes during cell division, the sister chromatids are held together at multiple sites from the time of their duplication. This sister chromatid cohesion is maintained by special protein molecules called the cohesin complexes. At the time when the chromatids of the chromosome pairs are being pulled to opposite poles of the dividing cell, the cohesin complexes connecting the sisters are degraded by a special protease called the separase. Thus chromosome segregation requires the timely destruction of the cohesin complexes as well as the removal of the DNA intertwines. For reviews, see Nasmyth, K. 2005. *Annu. Rev. Biochem.* **74:** 595–648; Yanagida, M. 2005. *Philos. Trans. R. Soc. B* **360:** 609–621; Hirano, T. 2006. *Nat. Rev. Mol. Cell Biol.* **7:** 311–322.

26. Holm, C., et al. 1989. *Mol. Cell. Biol.* **9:** 159–168.

A Checkpoint for Chromosome Decatenation?

It is well known that, in eukaryotes in particular, there are cell-cycle checkpoints to ensure the production of faithful replicas of the cells. These checkpoints monitor whether a cell in G_1 is ready for entry into the S phase, whether the duplicated DNA is in need of repair before a cell enters the M phase, whether the chromosome pairs are properly aligned at the mitotic plate for partitioning into the progeny cells, etc. Cell-cycle progression from one stage to the next is arrested or delayed if any imperfection is detected, to allow the cell time to remedy the problem.

As we have just seen, it appears that the progression of the cell cycle in yeasts is not significantly hindered by the absence of a functional type II DNA topoisomerase. A pair of dividing cells with intertwined sister chromatid pairs would apparently pass through G_2 and M and suffer a high probability of chromosome loss and/or chromosome breakage during mitosis. Since the mid-1990s, however, studies in mammalian cells suggested that there might be a Top2-related G_2 checkpoint, which would become activated by Top2 inhibition.[27]

The existence of a plausible cell-cycle checkpoint that senses incomplete decatenation between different chromosomes is surprising. Unlike cellular defects such as the presence of breaks or single-stranded regions in a DNA, it is unclear how the presence of a low degree of catenation between chromosomes could show a clear chemical signature. One suggestion is that the G_2 decatenation checkpoint might be triggered at the step of "chromosome individualization," that is, the step at which each chromatid pair of a cell condenses into a separate entity.[28] If two sister chromatid pairs happen to be tangled owing to the absence of Top2, then the tension between them, developed at this step by the compaction process, might trigger this G_2 decatenation checkpoint.[28] Because all studies thus far on the G_2 decatenation checkpoint in mammalian cells have used a bisdioxopiperazine compound to inhibit the type II DNA topoisomerases, the possibility remains that the observed checkpoint is a response to the trapping of Top2 molecules on the DNA, rather than the presence of linked chromosomes *per se* (as we will see in Chapter 10, an inhibitor of the bisdioxopiperazine class of drugs can trap a DNA-bound Top2 molecule on the DNA by locking its N-gate).

ROLE OF DNA TOPOISOMERASES IN CHROMOSOME CONDENSATION

Inside a cell, the DNA is organized into various compact forms by special proteins. In eukaryotes, chromosome compaction begins with nucleosome formation. In a nu-

27. Downes, C.S., et al. 1994. *Nature* **372:** 467–470 (Erratum in: 1994. *Nature* **372:** 710); for reviews, see Clarke, D.J., et al. 2006. *Cell Cycle* **5:** 1925–1983; Damelin, M. and Bestor, T.H. 2007. *Brit. J. Cancer* **96:** 201–205.

28. Giménez-Abián, J.F., et al. 2000. *Chromosoma* **109:** 235–244.

cleosome core particle, a stretch of about 145 bp of DNA makes two left-handed turns around a protein core that contains two each of the histones H2A, H2B, H3, and H4.[29] A string of nucleosomes on a DNA can fold further into higher order structures of increasing compactness in the presence of other proteins, culminating in the formation of a highly condensed structure at the time of mitosis. Relative to a naked DNA itself, a mitotic chromosome has a linear compaction ratio of hundreds to tens of thousands in different organisms, and, on staining with an appropriate reagent, is readily visible under an optical microscope. The degree of compaction of a mitotic chromosome relative to the same chromosome outside the M phase is more modest, with a linear compaction ratio varying from several to ~50-fold. Although the precise structure of a mitotic chromosome remains a subject of intense studies, all present models involve various levels of coiling of the DNA, beyond its wrapping around a nucleosome core. With all this coiling in the organization of intracellular DNA in general and the compaction of mitotic chromosomes in particular, entanglement within a single chromosome, or between different chromosomes, would seem inevitable in the absence of appropriate DNA topoisomerases.

In spite of a lack of studies of chromosome condensation in the absence of *all* DNA topoisomerases, there are ample hints in favor of the idea that at least one DNA topoisomerase is needed for the compaction of intracellular DNA. In the budding yeast, mitotic condensation of a particular chromosome region bearing a couple of hundred tandem copies of the ribosomal DNA genes appears to involve Top1.[30] There are also many experiments that implicate a requirement for a type II DNA topoisomerase at the late stage of mitotic chromosome condensation, even in the presence of the type I DNA topoisomerases.[31] In the *S. pombe top2 ts* mutant, for example, the final stage of chromosome condensation is blocked at restrictive temperatures.[23] Studies in other eukaryotes, using various DNA topoisomerase inhibitors, or reagents that interfere with the expression or the activity of the enzyme, led to essentially the same conclusion.[32] The use of mitotic cell extracts that promote chromosome condensation in vitro also shows that depletion of the type II DNA topoisomerase activity impinges on chromosome condensation.[33]

The involvement of the yeast type IB enzyme in ribosomal DNA (rDNA) condensation probably represents a special case, and its mechanism is further compli-

---

29. Kornberg, R.D. 1974. *Science* **184:** 868–871; Luger, K., et al. 1997. *Nature* **389:** 251–260.
30. Castano, I.B., et al. 1996. *Genes Dev.* **10:** 2564–2576.
31. For reviews, see Wang, J.C. 1996. *Annu. Rev. Biochem.* **65**: 635–692; Koshland, D. and Strunnikov, A. 1996. *Annu. Rev. Cell Dev. Biol.* **12:** 305–333; Belmont, A.S. 2006. *Curr. Opin. Cell Biol.* **18:** 1–7.
32. See, for example, Chang, C.J., et al. 2003. *J. Cell Sci.* **116:** 4715–4726; Sakaguchi, A. and Kikuchi, A. 2004. *J. Cell Sci.* **117**: 1047–1054; Ishida, R., et al. 1994. *Cell Biol.* **126:** 1341–1351.
33. Adachi, Y., et al. 1991. *Cell* **64:** 137–148; Hirano, T. and Mitchison, T.J. 1993. *J. Cell Biol.* **120:** 601–612; Cuvier, O. and Hirano, T. 2003. *J. Cell Biol.* **160:** 645–655.

cated by the finding that the dependence on DNA topoisomerase I was observed only in the absence of the product of a gene called *TRF4*.[30] On the other hand, the requirement of a type II DNA topoisomerase in the late stage of mitotic chromosome condensation seems quite general. How do the type IB and type II enzymes differ in solving the entanglement problems accompanying chromosome condensation? Why is a type II enzyme necessary in the presence of the type IB enzyme?

There are several plausible explanations. First, although both type IB and type II DNA topoisomerases can facilitate the folding of a long DNA when this folding can be accomplished by rotating the DNA around its helical axis, only a type II can accomplish the linking (catenation) or unlinking (decatenation) of DNA loops. Thus the role of the type II enzyme in late-stage mitotic condensation might be the removal of links between chromosomal loops that became entrapped during the folding of a chromosome. As we noted in the section on the Top2-dependent G_2 checkpoint, chromosome compaction may require their disentanglement. Second, the ATPase-coupled ability of a type II enzyme to bring two DNA segments together might serve as a facilitator of loop formation in chromatin; a chromatin loop transiently held together by a type II DNA topoisomerase could then be stabilized by other proteins involved in chromosome condensation. Third, several studies have led to the idea that mammalian Top2α may play a structural role in mitotic condensation. In one model, the string of nucleosomes in a chromosome was thought to first coil up into a filament 30 nm in diameter. This 30-nm fiber would then fold into multiple loops that become anchored to an axial scaffold of the chromosome, in which Top2α had been implicated as a major component.[34] Earlier experiments on a role of Top2 in chromosome scaffolding, using chemically fixed metaphase chromosomes,[35] were subsequently supported by viewing in a fluorescence microscope, living cells expressing a recombinant Top2α fused to a fluorescent protein.[36]

Nevertheless, the functional importance of the association between Top2α and the chromosome axial scaffold remains uncertain, and the known properties of the enzyme also provide little clue on how the enzyme might participate in attaching chromosome loops to a scaffold. Unless the entrance gate of the type II topoisomerase (discussed in Chapter 5) is somehow locked up after its binding to DNA (for example, by association of its ATPase domains with another protein), the binding of Top2 to DNA is not particularly tight. Inside a cell, the form of Top2 associated with condensed chromosomes appears to exchange readily with the unbound form.[36,37] Such a mobile

34. Marsden, M.P. and Laemmli, U.K. 1979. *Cell* **17:** 849–858; Maeshima, K., et al. 2005. *Chromosoma* **114:** 365–375. For reviews, see Losada, A. and Hirano, T. 2001. *BioEssays* **23:** 924–936; Gassmann, R., et al. 2004. *Expt. Cell Res.* **296:** 35–42; Belmont, A.S. 2006. *Curr. Opin. Cell Biol.* **18:** 1–7.

35. Earnshaw, W.C., et al. 1986. *J. Cell Biol.* **100:** 1706–1715; Gasser, S.M., et al. 1986. *J. Mol. Biol.* **188:** 613–629.

36. Tavormina, P.A., et al. 2002. *J. Cell Biol.* **158:** 23–29.

37. Christensen, M.O., et al. 2002. *J. Cell Biol.* **157:** 31–44.

association between the topoisomerase and DNA is contrary to what might be expected of a protein with a structural role in fastening the DNA loops to a scaffold. Furthermore, although a type II DNA topoisomerase is required for chromosome condensation in mitotic extracts of frog eggs, after condensation the enzyme can be readily removed from the condensed chromosomes without significantly affecting their morphology.[38] These observations raise questions as to whether Top2 is an integral component of a structurally important chromosome scaffold.

The three possibilities described above are not mutually exclusive, however, and only future experiments can tell us whether any one of them is valid. The plausible involvement of one or more DNA topoisomerases in the reversal of chromosome condensation, that is, chromosome decondensation or the unfolding of a condensed chromosome into a more open structure, is also in need of experimental studies.

DNA TOPOISOMERASES IN RECOMBINATION AND REPAIR

A number of enzymes that are structurally and evolutionarily related to the DNA topoisomerases are known to participate in DNA recombination, that is, the exchange of DNA segments and hence genetic information, between chromosomes. This exchange can occur during meiosis when recombination is required for proper chromosome pairing. And, as we shall see, the processes of replication and recombination share further common features. In sexually reproducing organisms, meiotic recombination between the paternal and maternal partner of a chromosome pair is initiated by the cleavage of double-stranded DNA by a type II DNA topoisomerase-related protein, as exemplified by the budding yeast *SPO11* gene product.[39] It remains one of Nature's secrets why a topoisomerase-like activity rather than a nuclease is used to initiate a double-stranded DNA break during meiotic recombination. We discussed also, in Chapter 4, the similarity between type IB DNA topoisomerases and the tyrosine recombinases that participate in site-specific recombination.

Are the archetype DNA topoisomerases involved in DNA recombination? There is strong evidence that they do, especially in recombinational repair, but the nature of their involvement may be different from that of the Spo11-like proteins or the tyrosine recombinases. A case in point is the plausible roles of the type IA subfamily of enzymes in recombination, which we shall consider in the remainder of this chapter.

38. Hirano, T. and Mitchison, T.J. *J. Cell Biol.* **120:** 601–612.

39. Bergerat, A., et al. 1997. *Nature* **386:** 414–417; Keeney, S., et al. 1997. *Cell* **88:** 375–384; Nichols, M.D., et al. 1999. *EMBO J.* **18:** 6177–6188; for reviews, see Keeney, S. 2001. *Curr. Top. Dev. Biol.* **52:** 1–53; Keeney, S. and Neale, M.J. 2006. *Biochem. Soc. Trans.* **34:** 523–525.

All Hell Breaks Loose on Inactivation of Top3 in Budding Yeast Cells

The requirement for at least one type IA DNA topoisomerase in a diverse spectrum of living organisms ranging from *E. coli* to mouse suggests that the enzyme plays a key cellular role. A number of studies indicate that one important function of the type IA enzymes is probably the resolution of a DNA structure or structures that are formed during recombination.

Because there is only a single type IA enzyme in the yeasts, functional studies of this enzyme in the budding yeast *S. cerevisiae* and the fission yeast *S. pombe* have been particularly revealing.[40–43] Although deletion of the *TOP3* gene (denoted *top3*Δ) is lethal in *S. pombe* as well as a number of *S. cerevisiae* strains,[43] *top3*Δ cells of at least one widely used *S. cerevisiae* strain are still capable of propagating, albeit poorly, and of forming small colonies on agar plates containing the proper nutrients.[40] This poor growth phenotype appears to reflect a retarded cell cycle. The *top3*Δ cells are often delayed in the late S/G2 phase, and, when viewed under a light microscope, are seen as large budded cells with an undivided nucleus at the neck of the bud.[40,41] These viable mutant cells also show a high recombination rate between repetitive nucleotide sequences on the same or different chromosomes, high sensitivity to DNA damaging agents, and, in the case of diploid cells, a defect in forming viable spores when the cells undergo sporulation.[40,42]

Tracing the Complex Phenotype of Budding Yeast *top3*Δ Cells to Defective Recombinational Repair

What might be the mechanism underlying the complex phenotype of the budding yeast *top3*Δ cells described previously? One powerful approach in understanding the phenotype of a mutant is by identifying second-site mutations that counter the effects of the original mutation. Second-site "suppressor mutations" can occur either in the same gene bearing the original mutation ("intragenic suppressors") or in different genes ("extragenic suppressors"). The latter class is particularly useful in identifying the other genes that act in the same pathway as the gene bearing the original mutation.

---

40. Wallis, J.W., et al. 1989. *Cell* **58:** 409–419.
41. Gangloff, S., et al. 1994. *Mol. Cell. Biol.* **14:** 8391–8398.
42. Watt, P.M., et al. 1996. *Genetics* **144:** 935–945; Yamagata, K., et al. 1998. *Proc. Natl. Acad. Sci.* **95:** 8733–8738; Frei, C. and Gasser, S.M. 2000. *J. Cell Sci.* **113:** 2641–2646; Saffi, J., et al. 2000. *Curr. Genet.* **37:** 75–78; Mullen, J.R., et al. 2000. *Genetics* **154:** 1101–1114; Chakraverty, R.K., et al. 2001. *Mol. Cell Biol.* **21:** 7150–7162.
43. Goodwin, A., et al. 1999. *Nucleic Acids Res.* **27:** 4050–4058; Onodera, R., et al. 2002. *Genes Genet. Syst.* **77:** 11–21.

The first such screen for suppressors of the slow-growth phenotype of the budding yeast *top3* mutants netted a range of second-site mutants, all of which map to an *S. cerevisiae* gene that became known as *SGS1* (*SGS* stands for slow growth suppression).[40] Mutations in *SGS1* suppress most phenotypic changes in *top3* mutants, including slow growth, although the meiotic defect of *top3* diploids in producing viable spore is not much helped by mutating *SGS1*.[40–44] Two aspects of the *SGS1* gene are particularly informative and have attracted much attention. First, its nucleotide sequence suggests that it may encode a DNA helicase, an enzyme that uses ATP hydrolysis to facilitate the separation of the base-paired strands in a DNA double helix. Indeed, the purified protein Sgs1 has been shown to possess a DNA helicase activity capable of unwinding duplex DNA molecules from their ends.[45] The identification of Sgs1 as a DNA helicase is revealing because a functional association between a type IA DNA topoisomerase and a helicase-like activity had been suggested: The enzyme "reverse gyrase" (described in Chapter 3) contains a type IA DNA topoisomerase domain and a DNA helicase-like domain in a single polypeptide. The parallel between the Top3-Sgs1 pair and reverse gyrase is further underscored by evidence that Top3 physically and functionally interacts with Sgs1.[41,43,46]

Second, the amino acid sequence of Sgs1 indicates that it is a homolog of the *E. coli recQ* gene product. Because *E. coli recQ* has been implicated in one of the homologous recombination pathways,[47] the involvement of a *recQ* homolog as a *top3* suppressor is suggestive that the type IA DNA topoisomerase might also participate in homologous recombination.

It turns out that two of the five known human *RECQ* homologs are linked to two distinct genetic diseases that are characterized by genome instability and predisposition to different forms of cancer: the Bloom syndrome and the Werner syndrome.[48,49] These genes (both *RECQ* homologs) have therefore been denoted *BLM* and *WRN*, respectively. A third *RECQ* homolog, *RECQ4*, has been linked to a subset of the Rothmund–Thomson syndrome, a heritable disease also characterized by genome instability and cancer susceptibility.[50] The finding of a link between Sgs1-like proteins and several human genome instability syndromes has stimulated a great deal of interest in Sgs1, as well as how the RecQ helicases are functionally related to the type IA DNA topoisomerases.

44. Maftahi, M., et al. 1999. *Nucleic Acids Res.* **27:** 4715–4772.

45. Bennett, R.J., et al. 1999. *J. Biol. Chem.* **273:** 9644–9650; Bennett, R.J., et al. 1999. *J. Mol. Biol.* **289:** 235–248.

46. Bennett, R.J., et al. 2000. *J. Biol. Chem.* **275:** 26898–26905.

47. For reviews, see Kowalczykowski, S.C. 2000. *Trends Biochem. Sci.* **25:** 156–65; Heyer, W.D. 2004. *Curr. Biol.* **14:** R895–897; Nakayama, H. 2005. *Mutat. Res.* **577:** 228–236.

48. Ellis, N.A., et al. 1995. *Cell* **83:** 655–666.

49. Yu, C.E., et al. 1996. *Science* **272:** 258–262; Kitao, S., et al. 1999. *Genomics* **61:** 268–276; Wang, L.L., et al. 2001. *Am. J. Med. Gene.t* **102:** 11–17.

50. Shor, E., et al. 2002. *Genetics* **162:** 647–662.

In addition to *SGS1*, a number of additional suppressor genes of budding yeast *top3* mutants have been identified. These include *RAD51*, *RAD54*, *RAD55*, and *RAD57*, all of which are known to be involved in homologous recombination.[50,51] The identification of various *top3*Δ suppressor genes that normally function in homologous recombination is consistent with the idea, first suggested by Rodney Rothstein and his associates, that at least one type IA DNA topoisomerase is required for the proper resolution of a DNA structure or structures that are formed in a particular homologous recombination pathway.[41] According to this interpretation, if the particular recombination pathway is blocked by a mutation in one of the *top3*Δ suppressor genes, then the DNA structure or structures that require a type IA DNA to resolve would not form, and there would be no need for the particular topoisomerase. A corollary of this postulate is that Top3 itself is not required in forming the recombination structure or structures in this particular pathway; otherwise, in a *top3* mutant no such a recombination intermediate or intermediates would form whether Sgs1 is present or not.

Further support of this idea just described comes from a study of the meiotic defect of yeast *top3* diploid cells. Yeast cells lacking DNA topoisomerase III can apparently enter meiosis and initiate the Spo11-mediated recombination between paternal and maternal chromosomes, but they cannot complete the first meiotic cell division. If recombination between homologous chromosome pairs is prevented from occurring by deletion of the *SPO11* gene, however, then diploid *top3*Δ cells become capable of yielding viable spores.[52] In other words, Top3 is dispensable in sporulation if meiotic recombination is bypassed.

From Yeast to Human

Despite the myriad forms of living organisms, many fundamental processes in different living cells are remarkably similar. Thus ideas that emerged from studies of the physiological roles of the sole type IA DNA topoisomerase of the budding yeast *S. cerevisiae* are likely to be applicable to this subfamily of enzymes in other living organisms. For example, the fission yeast *S. pombe* and the budding yeast *S. cerevisiae* diverged from one other hundreds of million years ago, but the consequences of inactivating Top3 in the two organisms are rather similar. Inactivation of *S. pombe* Top3 also leads to mitotic chromosome missegregation and hypersensitivity to DNA damaging agents.[42–44,53] Furthermore, *S. pombe top3*Δ cell lethality can also be suppressed by mutations in its *recQ*

---

51. Oakley, T.J., et al. 2002. *DNA Repair* **1:** 463–482.

52. Gangloff, S., et al. 1999. *EMBO J.* **18:** 1701–1711. The experiments were performed in a *spo13*Δ genetic background. During sporulation of a *spo13*Δ diploid cell, the first meiotic division is bypassed, and the chromosomes of a diploid cell duplicate once before cell division to yield two diploid spores. Inactivation of DNA topoisomerase III in the *spo13*Δ background abolishes viable spore formation, unless the Spo11 protein is also inactivated.

53. Laursen, L.V., et al. 2003. *Mol. Cell Biol.* **273:** 3692–3705.

homolog gene *rqh1* (also termed *rec12*) as well as several other genes involved in homologous recombination.[53–55] Similar to the budding yeast RecQ homolog Sgs1, the fission yeast Rqh1 also physically interacts with Top3 proteins.[53,55]

In contrast to the presence of a single type IA DNA topoisomerase in yeasts, in bacteria such as *E. coli* there are two type IA DNA topoisomerases I and III, encoded, respectively, by the genes *topA* and *topB*. Elimination of both genes leads to aberrant chromosome morphology and is lethal[56,57]; this lethality has similarly been attributed to a requirement of at least one of the type IA enzymes in a homologous recombination pathway.[56] Vertebrates also have two type IA enzymes DNA topoisomerases IIIα (Top3α) and IIIβ (Top3β), encoded, respectively, by the genes *TOP3*α and *TOP3*β. Human Top3α has been shown to interact with the Blm and RecQL1 proteins, two of the five human RecQ homologs.[58] Functional and physical interaction between Top3α and Blm in an avian cell line has also been reported.[59]

As we discussed at the beginning of this chapter, inactivation of mouse *TOP3*α is lethal, and the *top3*$\alpha^{-/-}$ embryos die around the time of their implantation into the uterus wall.[4] Although *top3*$\beta^{-/-}$ mice are viable and develop normally, their fecundity progressively decreases over time and from one generation to the next, and their lifespan is only about one-half as long as that of their *TOP3*β^{+} siblings owing to the development of autoimmunity as the mutant animals age.[6] Whereas tracing the molecular root of a particular phenotype of an organism as complex as a mouse is a challenging task, the phenotypes of the *top3*$\alpha^{-/-}$ and *top3*$\beta^{-/-}$ mutants can also be attributed to a critical role of a type IA DNA topoisomerase in the proper segregation of chromosomes after meiotic or mitotic recombination. In fact, germ cells as well as somatic cells from *top3*$\beta^{-/-}$ mice show a high frequency of abnormal chromosome numbers, consistent with aberrant chromosome segregation in these cells.[6]

Plausible DNA Structures that May Require a Type IA DNA Topoisomerase in Their Processing

The results discussed above hint at a requirement for a type IA DNA topoisomerase in processing the product(s) of a certain homologous recombination pathway or pathways, and that improper processing of such product(s) could lead to an accumulation of DNA damage and chromosome missegregation during cell division. The exact nature of these putative intermediates remains a matter of speculation, however.

54. Oh, M., et al. 2002. *Nucleic Acids Res.* **30:** 4022–4031; Win, T.Z., et al. 2004. *J. Cell Sci.* **117:** 4769–4778.
55. Ahmad, F. and Stewart, E. 2005. *Mol. Genet. Genomics* **273:** 102–114.
56. Zhu, Q., et al. 2001. *Proc. Natl. Acad. Sci.* **98:** 9766–9771.
57. Stupina, V.A. and Wang, J.C. *J. Biol. Chem.* **280:** 355–360.
58. Wu, L., et al. 2000. *J. Biol. Chem.* **275:** 9636–9644; Johnson, F.B., et al. 2000. *Cancer Res.* **60:** 1162–1167.
59. Seki, M., et al. 2006. *Mol. Cell Biol.* **26:** 6299–6307.

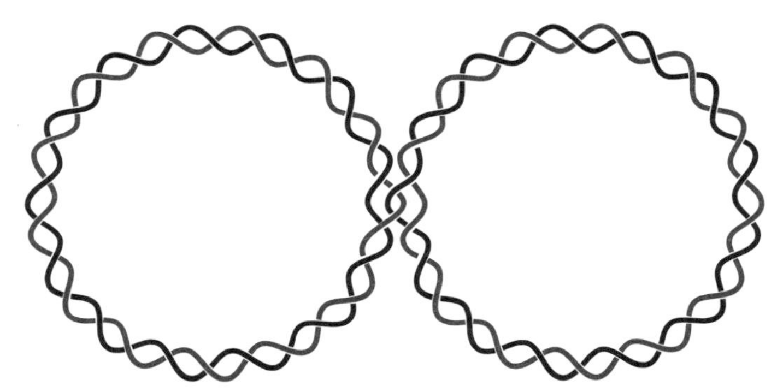

Figure 7-6. A hemicatenane in which two strands of a pair of duplex DNA rings are linked, either singly as depicted, or multiply.

To gain some insight into this problem, it is useful to consider some of the structural aspects of recombination intermediates. From a structural perspective, the replication and recombination of a DNA molecule share many common features. The replication intermediate shown in the rightmost panel in Figure 7-4, for example, can also represent a recombination intermediate that is formed by intertwining complementary single-stranded regions in a pair of gapped DNA duplexes. This structure is one of a plethora of structures in which two DNA molecules are topologically linked by intertwined strands. The structures shown in Figures 7-6 and 7-7 illustrate two other examples that have attracted considerable attention in discussions on recombination in general and on the role of the type IA DNA topoisomerases in recombina-

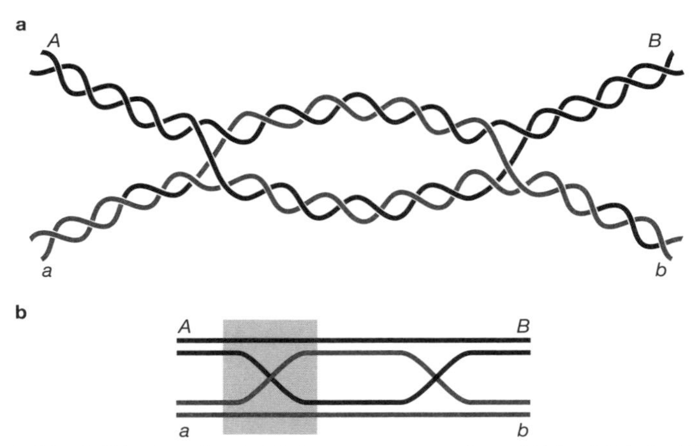

Figure 7-7. A double Holliday junction (DHJ) in two representations. (*a*) The DNA double helix is represented by a pair of intertwined strands. (*b*) Here, the DNA double helix is represented by a pair of parallel lines. The structure inside the shaded box is termed a Holliday junction. A/a and B/b denote allele pairs on either side of the junction (an allelic pair refers to two alternative forms of the same gene).

tion in particular. The structure shown in Figure 7-6 is termed a hemicatenane; it contains two DNA rings that are topologically linked by intertwines between two DNA strands from two separate double-stranded DNA molecules. Despite its name, this class of joined molecules is not limited to DNA rings; such a linked structure can form between linear molecules as well. DNA hemicatenane was first identified in a reaction catalyzed by the *E. coli* RecA protein between a double-stranded and a single-stranded DNA ring that share regions of sequence homology,[60] but such structures have also been implicated in replication-dependent processes.[61]

The structure shown in Figure 7-7, in which the DNA double helix is depicted as a pair of intertwined strands in Figure 7-7a or more simply as two parallel lines in Figure 7-7b, is termed a double Holliday junction (DHJ). It is so named because it constitutes two Holliday junctions, a structure that was first proposed in 1964 by Robin Holliday as a key recombination intermediate (the structure inside the shaded box shown in Fig. 7-7b).[62] The structure, formation, and resolution of the Holliday junction (HJ) have been discussed in detail in biology and genetics textbooks, but these discussions are only peripherally related to the issues addressed in this chapter and so are largely omitted here. The formation of the DHJ has also been implicated in a number of models for recombination,[63] and there is substantial evidence for its formation during meiotic recombination in the budding but not the fission yeast.[64] In some recombination pathways, DHJ formation is probably preceded by the formation of a double-stranded break in one of the pair of DNA molecules undergoing recombination, as illustrated in Figure 7-8. In principle, however, DHJ formation can also occur between intact DNA strands through the action of a type I DNA topoisomerase, as first pointed out by J. J. Champoux in 1977 (Fig. 7-9).[65] Both HJ and DHJ formation are often implicated in the repair of replication forks that become stalled or are broken when the replication machinery encounters a damaged site in a DNA (Fig. 7-10).[66]

Why Can the Role of a Type IA DNA Topoisomerase Not Be Substituted by Other Enzymes?

In all of the structures discussed previously, pairs of DNA molecules are topologically joined by intertwine between two DNA strands. As mentioned earlier in this chapter, a type IA DNA topoisomerase *E. coli* DNA topoisomerase III has been shown to per-

60. Cunningham, R.P., et al. 1981. *Cell* **24:** 213–223.

61. Sogo, J.M., et al. 1986. *J. Mol. Biol.* **189:** 189–201; Lucas, I. and Hyrien, O. 2000. *Nucleic Acid Res.* **28:** 2187–2193; Lopes, M., et al. 2003. *J. Mol. Cell* **12:** 1499–1510.

62. Holliday, R. 1964. *Genet. Res.* **5:** 282–304.

63. See, for example, Szostak, J.W., et al. 1983. *Cell* **33:** 25–35.

64. Schwaca, A. and Kleckner, N. 1995. *Cell* **83:** 783–791; Cromie, G.A., et al. 2006. *Cell* **127:** 1167–1178.

65. Champoux, J.J. 1977. *Proc. Natl. Acad. Sci.* **74:** 5328–5332.

66. See, for example, Kuzminov, A. 2001. *Proc. Natl. Acad. Sci.* **98:** 8461–8468.

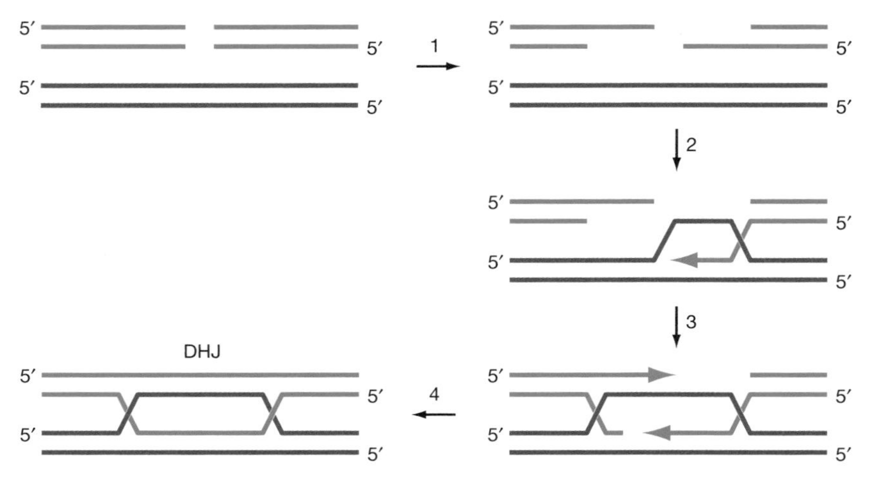

Figure 7-8. A model in which recombination between a pair of DNA molecules of identical or nearly identical nucleotide sequence (homologous recombination) is initiated by a double-stranded break in one of the pair (Szostak, J.W., et al. 1983. *Cell* **33:** 25–35). One DNA molecule is represented by a pair of blue lines; in the other DNA molecule (pair of red lines), a double-stranded break has occurred. In Step *1*, specific enzymes involved in homologous recombination resect the broken DNA to expose single-stranded 3′ ends. In Step *2*, the recombination machinery mediates the invasion of the intact DNA by a 3′ single-stranded end to form a structure with a single-stranded loop (a "D-loop" structure). In Step *3*, extension of the invading 3′ end by repair synthesis lengthens the single-stranded loop, and this exposed region in turn allows its pairing with the other broken end. In Step *4*, a DHJ is formed by repairing the gaps through repair synthesis and sealing the nicks by DNA ligase.

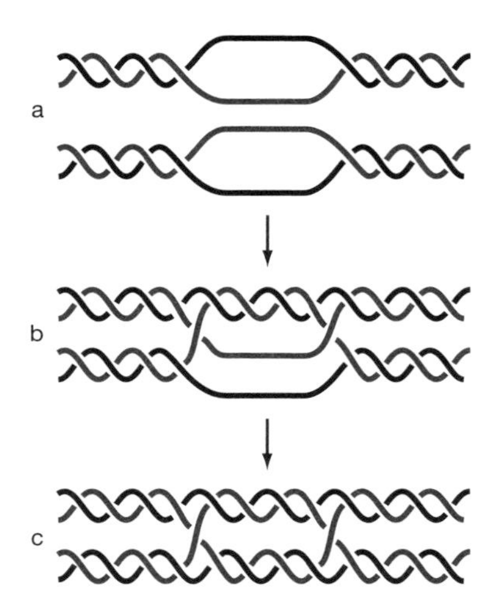

Figure 7-9. Formation of a double Holliday junction by a type I DNA topoisomerase. A pair of DNA molecules with homologous nucleotide sequences are shown to form single-stranded loops at corresponding positions. Sequential pairing and intertwining of the two pairs of strands with complementary nucleotide sequences in the presence of a type I DNA topoisomerase lead to the formation of a DHJ. (Redrawn, with permission, from Champoux, J.J. 1977. *Proc. Natl. Acad. Sci.* **74:** 5328–5332.)

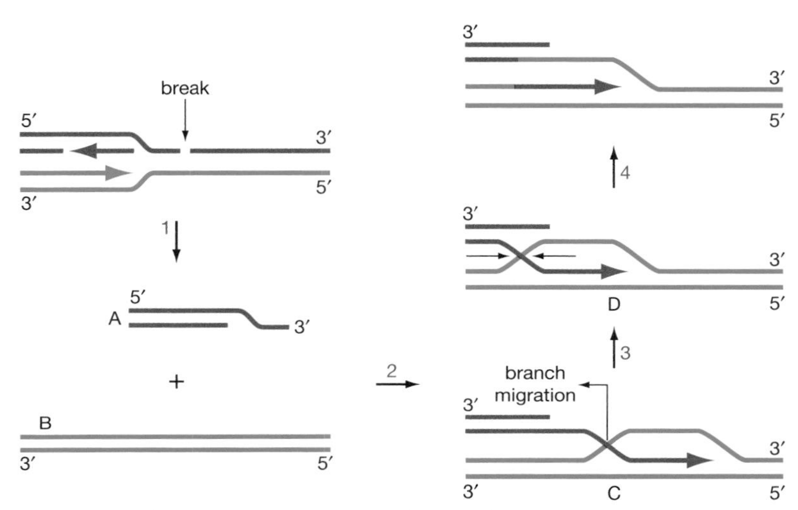

Figure 7-10. An example of replication fork repair by homologous recombination. The fork encounters a single-stranded break in the parental DNA and breaks apart, but elongation of one of the strands (the leading strand shown in the drawing) can continue past the location of the original break to yield the pair of products A and B (Step *1*). The repair system can then mediate the invasion of B by the 3′ end of A to give the structure C (Step *2*). The strand crossing shown in C can move relative to the DNA ends, either to the right or left, in a process termed "branch migration" (Step *3*): Disruption of base pairs between the red strands to the left of the crossing concurrently with base pairing between the red and blue strand to the right of the crossing would result in leftward movement of the crossing, and disruption in the other direction would result in rightward movement of the crossing. Branch migration can occur in a structure like C by thermal motion, or can be driven by a protein machinery. The structure D formed by branch migration contains a Holliday junction. The replication fork is restored by cutting the crossing strands in D at the positions indicated by the pair of arrows, followed by repair of the cut strands (Step *4*).

mit the *complete* separation of a pair of intertwined DNA strands in a purified plamid DNA replication system.[12] It has also been shown that a DHJ can be completely resolved by the combined actions of human Top3α and the Blm helicase, or their *Drosophila* counterparts.[67]

There are often, however, other pathways for resolving the same DNA structures without the participation of a type IA enzyme. For example, as illustrated in Fig. 7-5, a pair of intertwined DNA strands can also be resolved by a type II DNA topoisomerase after its conversion to a pair of intertwined DNA double helices. Furthermore, the combined actions of DNases and other enzymes provide additional ways of resolving entangled DNA strands.

67. Wu, L. and Hickson, I.D. 2006. *Nature* **426:** 870–874; Plank, J.L., et al. 2006. *Proc. Natl. Acad. Sci.* **103:** 11118–11123.

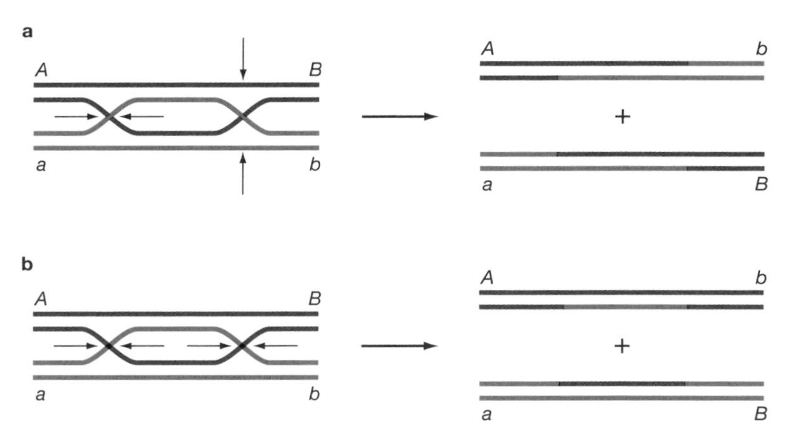

Figure 7-11. Cutting a double Holliday junction (DHJ). The pair of small arrows in each drawing indicates the positions where breakage of the DNA strands is made to resolve the pair of DNA molecules (*a*) Here a pair of recombinant DNA molecules is formed. (*b*) In this scheme a pair of nonrecombinant DNA molecules is formed.

In some situations, however, pathways using a type IA enzyme appear to offer a unique advantage. One example is the resolution of a DHJ. A DHJ can be resolved in two ways: either by nucleolytic cleavage of DNA strands followed eventually by rejoining of the broken strands, or by the use of a DNA topoisomerase to carry out both strand cleavage and rejoining. If a DHJ is resolved by nucleolytic cleavage, two different classes of products are plausible depending on which pair of strands is cleaved at each HJ. For the DHJ illustrated in Figure 7-11, cleaving the inside pair of strands at one HJ and the outside pair of strands at the other HJ (positions of cleavage indicated by arrows in Fig. 7-11a), or vice versa, would give products in which genes on the left side of these products would be derived from one parent DNA, and those on the right side of the products would be derived from another parent DNA molecule. This crossing-over of the flanking genes encoded by the two DNA molecules thus produces a pair of recombinants. On the other hand, if only the inside or the outside strands at both HJ are cleaved (positions of cleavage indicated by arrows in Fig. 7-11b), then the recombination products are nonrecombinants with respect to the flanking genes, and there is only exchange of DNA strands in the heteroduplex region. Here the "inside strand" and "outside strand" descriptions refer strictly to the particular drawings shown, and the two can readily interconvert when a HJ is shown in an open form, in which the "inside" and "outside" strands are equivalent. (The actual geometry assumed by a particular HJ is determined by the sequences of the DNA strands forming the junction, as well as the divalent metal ions and specific proteins that bind HJ structures.)

However, if a DHJ is resolved by the topoisomerase-mediated removal of strand intertwines between the "heteroduplex" regions (so named because the two complementary strands in a heteroduplex originate from different members of the DNA pair), then only nonrecombinant DNA molecules are produced. This is because a DNA topoisomerase breaks and rejoins the *same* DNA strand, and thus there can be no swapping of strands with flanking genetic markers between the pair of DNA molecules forming the DHJ. Resolving a recombination intermediate without producing a recombinant is clearly advantageous in mitotic cells, in which the primary role of homologous recombination is to repair DNA lesions. Furthermore, among all DNA topoisomerases, a type IA DNA topoisomerase appears to be uniquely suited for the resolution of a DHJ. Unlike the branched structure shown in the right drawing in Figure 7-4, the heteroduplex intertwines in a DHJ cannot be readily converted to intertwines between double-stranded DNA segments for untanglement by a type II DNA topoisomerase. A type IB enzyme is probably also inadequate in removing the strand intertwines, because the action of a type IB enzyme requires it to interact with a double helix DNA segment, which makes the removal of the last couple of turns between two intertwined strands difficult. Thus for a structure like a DHJ, the complete separation of the two intertwined pair of DNA molecules may specifically require a type IA DNA topoisomerase, and neither a type II or a type IB enzyme may substitute for the type IA enzyme.[68]

Whereas a recombination intermediate like DHJ can be resolved by the combined action of Sgs1 and Top3, it is less clear why the formation of such a structure should depend on Sgs1. It is plausible that, similar to the role of RecA described in the previous section, Sgs1 and the other proteins identified in the suppressor screens are involved in opening up a recipient double-stranded DNA for pairing with an invading homologous single stranded DNA. It may also be significant that in the case of reverse gyrase, a helicase-type activity might augment a type IA activity in intertwining complementary strands.[69] Further biochemical studies are needed, however, to elucidate the precise molecular roles of the DNA topoisomerases in recombination and repair.

68. Wang, J.C. 2002. *Nat. Rev. Mol. Cell. Biol.* **3:** 430–440.

69. Hsieh, T.S. and Plank, J.L. 2006. *J. Biol. Chem.* **281:** 5640–5647.

CHAPTER 8

Manifestations of DNA Entanglement
Gene Expression

"From the Grand Void Ying and Yang simultaneously emerged." [1]

From The Book of Changes (I Ching or Yi Jing)

AMONG CELLULAR PROCESSES INVOLVING MACROMOLECULAR machineries moving along DNA, transcription is one of the best known and most extensively studied. In transcription, one of the two complementary DNA strands serves as the template for the synthesis of an RNA strand: The nucleotides of the template DNA strand are copied ("transcribed") by the enzyme RNA polymerase into RNA (ribonucleic acid). Chemically, RNA differs from DNA in two aspects. First, whereas in the DNA each nucleotide contains a deoxyribosyl sugar, in the RNA this moiety is a ribose with a hydroxyl group at its C_2' position (see Fig. 1-1b in Chapter 1 for a sketch of a DNA nucleotide and the numbering system; the C_2'-OH group in a ribose is pointing in a direction opposite to that of the base attached to the same sugar ring). Second, in RNA the base U (uracil) replaces the base T (thymine) in the corresponding DNA; the other bases are the same. U differs from a T by its lack of a 5-methyl group. During transcription, the nucleotides A, T, G, and C of the template DNA strand are respectively copied by the enzyme RNA polymerase into the ribose nucleotides U, A, C, and G in the RNA product, according to a copying scheme very similar to the base-pairing scheme between complementary DNA strands: A is copied into U, T to A, C to G, and G to C.

1. This liberal translation of a line from the first Xi Ci chapter of the ancient *Book of Changes* (*I Jing*) seems to be a fitting description of the "twin-supercoiled-domain" model described in this chapter.

The nascent RNA chain is extended by an RNA polymerase tracking along its DNA template, and the RNA remains attached to the polymerase during its synthesis, from the start to termination of the chain. Furthermore, inside a cell the nascent RNA is usually associated with other macromolecules, including the ribosomes in the case of a prokaryote, in which the translation of a polypeptide-coding RNA (a "messenger" RNA or mRNA) into the protein product often starts before the termination of the mRNA chain ("cotranscriptional translation"). Thus during transcription the RNA polymerase, the nascent RNA, and any other macromolecules associated with the nascent RNA constitute a large assembly that moves in a coordinated way along the DNA template. This assembly will be termed the "transcription assembly" and assigned a symbol R in subsequent discussions. In many instances, one or more components of R may be associated with a plethora of additional macromolecules, such as those involved in regulating transcription, in modifying the nucleoprotein structure of intracellular DNA, and in further processing of the RNA transcript, including its translation into protein.

DNA TOPOSIOMERASES AND TRANSCRIPTION

Moving along a Helical Path

It was realized long ago that an advancing RNA polymerase might have to revolve around its DNA template, owing to the helical geometry of DNA.[2] This relative rotation between a transcribing RNA polymerase and its DNA template resembles the rotation of a nut traversing a matching bolt, and has been directly shown (Fig. 8-1). In the single-molecule experiment illustrated in the figure, the transcribing *Escherichia coli* RNA polymerase was fixed to a glass surface, and the rotation of the DNA relative to the stationary polymerase was observed by viewing a paramagnetic bead, which was decorated with fluorescent markers, at one end of the DNA (see Chapter 6 for discussions on single-molecule experiments). During RNA synthesis under these conditions, it was observed the bead would rotate in a direction expected for a right-handed screw advancing through a stationary nut, and at an angular velocity consistent with the rate of RNA synthesis and the expectation that the bead would make a full rotation for every 10 DNA base pairs (bp) transcribed.[3]

This experiment illustrates that if a transcribing RNA polymerase is prevented from circling around the DNA (in this particular instance by fixing it to a glass surface), then the DNA template being transcribed would rotate around its own helical axis at a rate of about one turn per 10 bp transcribed. Would this transcription-driven axial rotation

2. Maaløe, O. and Kjeldgaard, N.O. 1966. *Control of macromolecular synthesis*. Benjamin, New York; Gamper, H.B. and Hearst, J.E. 1982. *Cell* **29:** 81–90.

3. Harada, Y., et al. 2001. *Nature* **409:** 113–115.

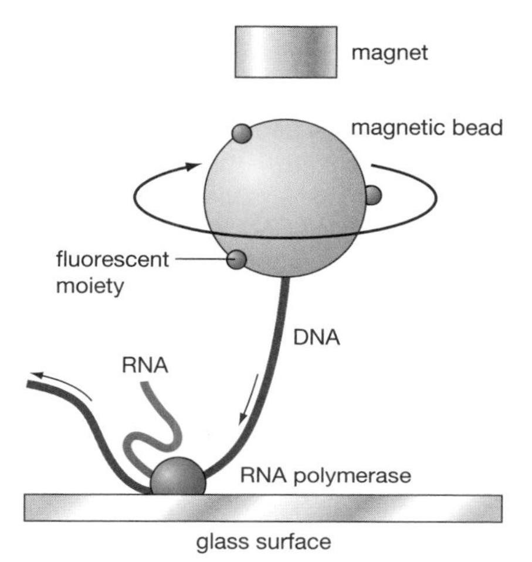

Figure 8-1. A single-molecule experiment demonstrating the relative rotation between a transcribing RNA polymerase and its DNA template. A magnetic bead was glued to one end of a 5-kb DNA bearing a strong promoter for phage T7 RNA polymerase. DNA was incubated with the phage polymerase in a transcription buffer lacking UTP to initiate the synthesis of a short transcript. (When the polymerase encounters the first A in the template, it is forced to pulse as no U is available.) The polymerase was then anchored to a glass slide, and a magnet was used to suspend the DNA-attached magnetic bead. UTP was then added to the transcription mixture to allow elongation of the arrested transcript, and rotation of the fluorescently decorated bead was monitored by viewing through a fluorescence microscope. (Redrawn, with permission of Macmillan Publishers Ltd., from Harada, Y., et al. 2001. *Nature* **409:** 113–115.)

of the DNA occur inside a cell, and what might be the consequences of this forced rotation of the DNA? We shall consider these issues in later sections of this chapter.

Transcription as a Driving Force for DNA Supercoiling inside Cells

The first hint that transcription may be closely related to the supercoiling of intracellular DNA came from findings reported in the mid-1980s by Gail J. Pruss and Karl Drlica. They observed that in *topA* mutants of *E. coli* or *Salmonella typhimurium* expressing a defective DNA topoisomerase I (Top1), a plasmid pBR322, but not its derivative pUC9, became twice as negatively supercoiled as the same plasmid isolated from a *topA*$^+$ strain with normal Top1 activity. Remarkably, this excessive negative supercoiling or "hypernegative supercoiling" of pBR322 in the absence of Top1 was found to depend on the transcription of the *tetA* gene carried on pBR322, which is absent in pUC9.[4] Expression of a functional *tetA* product, on the other hand, appeared unimportant for this phenomenon.[4] The striking dependence of the extent of

4. Pruss, G.J. 1985. *J. Mol. Biol.* **185:** 51–63; Pruss, G. and Drlica, K.1986. *Proc. Natl. Acad. Sci.* **83:** 8952–8956.

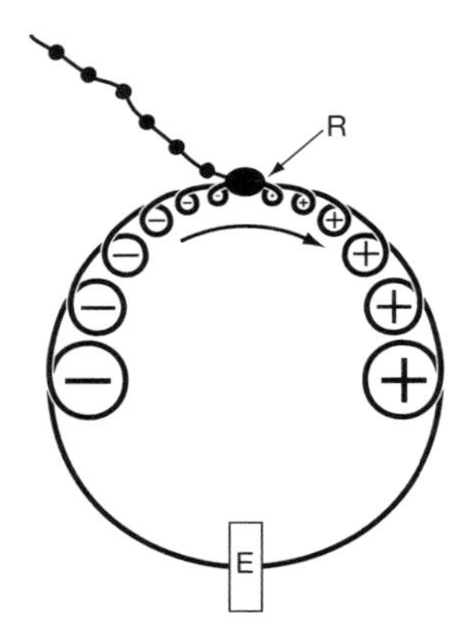

Figure 8-2. The twin-supercoiled-domain model. A transcription assembly R consisting of an RNA polymerase, the nascent RNA associated with the polymerase, and RNA-associated proteins is shown to move in the clockwise direction along a ring-shaped double-stranded DNA, here represented by a single line. The advancing R generates positive supercoils (+ signs) ahead of it and negative supercoils (– signs) behind it. A hypothetical element E, which represents a frictional barrier against the rotation of the DNA around its helical axis, is shown to divide the oppositely supercoiled regions. (Modified, with permission, from Wang, J.C. 1991. *J. Biol. Chem.* **266:** 6659–6662, ©American Society for Biochemistry and Molecular Biology.)

negative supercoiling of a plasmid on *tetA* transcription could not be explained by the prevailing model (at that time) of supercoiling regulation in bacteria. In this model, the extent of supercoiling of intracellular DNA is determined by a dynamic balance between the actions of gyrase and DNA topoisomerase I, the former negatively supercoils the DNA and the latter removes the negative supercoils.

The Twin-Supercoiled-Domain Model

The puzzling observation described in the previous section, that *tetA* transcription is causally related to excessive negative supercoiling of a *tetA*-bearing plasmid in *topA* mutants (and several other considerations), led to the proposal of the "twin-supercoiled-domain" model of transcription in 1987.[5] The model postulates that although inside a bacterium like *E. coli* or *S. typhimurium* a transcription assembly R can generally revolve around the template DNA, this is apparently untrue for the transcription of certain genes like *tetA*. Figure 8-2 illustrates this situation for a plasmid bearing a single transcription unit of the *tetA* category in which R is somehow prevented from revolving around the DNA template in the cellular milieu. Cranking the helical DNA through a transcribing R would then make the DNA region ahead of R more tightly wound or positively supercoiled—a situation very similar to the positive supercoiling of the DNA ahead of a replication fork (see Fig. 7-1 in Chapter 7). In the region behind R, the DNA becomes less tightly wound or negatively supercoiled as it is threaded through the immobile polymerase. Equal numbers of positive and negative supercoils are simultaneously generated by

5. Liu, L.F. and Wang, J.C. 1987. *Proc. Natl. Acad. Sci.* **84:** 7024–7027. Bacterial DNA topoisomerase IV was not yet discovered when the model was proposed, and gyrase was assumed to be the only enzyme capable of removing positive supercoils in the model; the presence of Top4 as an activity that augments gyrase in positive supercoil removal does not alter the major predictions of the model.

transcription when the RNA polymerase is prevented from circling its DNA template, and hence the name "twin-supercoiled-domain" model.

In the sketch shown in Figure 8-2, the two oppositely supercoiled domains are separated by R and a hypothetical element E, which, for the time being, can be regarded simply as a barrier against the cancellation of the positive and negative supercoils through axial rotation of the tethering DNA in between. A useful way of visualizing the generation of oppositely supercoiled domains by transcription is to consider both R and E as additional barriers to axial rotation of the DNA between oppositely supercoiled domains. In a DNA ring, axial rotation of either the tethering DNA bearing R, or that bearing E, would allow the cancellation of the positive and negative supercoils.

Pathways for the Removal of Positive and Negative Supercoils

In the twin-supercoiled-domain model, the removal of the supercoiled domains generated by transcription is thought to involve two entirely different pathways: axial rotation of the connecting DNA segments as described above, and the actions of the DNA topoisomerases. Because in the absence of the DNA topoisomerases the linking number of a DNA ring or loop is an invariant, the transcription process itself must generate exactly the same numbers of negative and positive supercoils, and exactly the same number of negative and positive supercoils must be removed by axial rotation of the DNA between two oppositely supercoiled domains.

Removal of the supercoils generated by transcription by the DNA topoisomerases presents a very different situation, however. This is because the linking number is no longer invariant in the presence of a topoisomerase, and the enzymes involved may have their intrinsic preferences for the removal of supercoils of different signs. As we described in Chapters 5 and 6, in bacteria like *E. coli* and *S. typhimurium*, both DNA gyrase and Top4 preferentially remove positive supercoils. Thus the action of these type II enzymes would favor a net accumulation of negative supercoils when equal numbers of positive and negative supercoils are continuously being generated by transcription. By contrast, the type IA enzyme bacterial DNA topoisomerase I specifically removes negative supercoils, and the rate of negative supercoil removal by this enzyme is steeply dependent on the degree of negative supercoiling—the reaction greatly slows down as the degree of negative supercoiling is lowered, and thus under normal cellular conditions a significant portion of the negative supercoils generated by transcription is likely to be retained in the DNA template. Another factor that may contribute to a net negative supercoiling of intracellular DNA in bacteria is the ability of DNA gyrase to actively drive DNA negative supercoiling: If the type II DNA topoisomerases are removing the positive supercoils faster than their generation by transcription, then gyrase would start to negatively supercoil the DNA template.

The twin-supercoiled-domain model and the intrinsic differences in the actions of the DNA topoisomerases readily account for the excessive negative supercoiling of a plasmid like pBR322 but not pUC9 in an *E. coli topA* mutant. First, the positive supercoils generated by transcription would be removed by DNA gyrase, and to a lesser extent Top4 (see the section "Entanglement of the DNA Strands during Replication" in Chapter 7), but the negative supercoils would accumulate in the absence of bacterial Top1. Thus in a *topA* mutant lacking Top1 the generation of equal number of positive and negative supercoils would be expected to lead to a much higher net accumulation of the negative supercoils. Second, for a plasmid like pUC9 lacking the *tetA* gene, the absence of Top1 does not lead to hypernegative supercoiling because the transcription assemblies on pUC9 are ineffective in generating oppositely supercoiled domains. A corollary of the twin-supercoiled-domain model of transcription is that bacterial DNA topoisomerases III and IV are normally not very robust in negative supercoil removal; otherwise inactivation of Top1 in the *topA* mutants would be insufficient for hypernegatively supercoiling.

The twin-supercoiled-domain model also provides an explanation of another earlier observation, that pBR322 becomes positively supercoiled in *E. coli* cells treated with a DNA gyrase inhibitor novobiocin.[6] This observation was initially very puzzling. If none of the known *E. coli* DNA topoisomerases can positively supercoil a DNA, how could a plasmid become positively supercoiled by inhibiting a topoisomerase or topoisomerases? The twin-supercoiled-domain model would predict, however, that transcription of the plasmid-borne *tetA* gene simultaneously generates both positive and negative supercoils in the plasmid, and thus a net accumulation of positive supercoils is expected when the negative supercoils are removed by Top1 whereas removal of positive supercoils is prevented by the presence of novobiocin (which, as it turns out, inhibits both DNA gyrase and Top4, the two enzymes capable of removing positive supercoils).

In eukaryotes and in bacteria that express a robust type IB DNA topoisomerase in addition to the type IA and IIA enzymes, however, it would be expected that both positive and negative supercoils could be removed by the type IB enzyme and Top2. Thus in these organisms it would be more difficult to show the generation of oppositely supercoiled domains through inhibition of one particular DNA topoisomerase.

What Prevents R from Revolving around DNA?

Experimental results since the initial proposal of the twin-supercoiled-domain model in 1987 have lent strong support to the model. But why is it the transcription of certain genes like *tetA*, rather than the transcription of any gene, that drives the supercoiling of the DNA template? What might be preventing a particular R from circling around the DNA template?

Several plausible explanations were suggested in the 1987 paper on transcriptional supercoiling, and two of the most significant are illustrated in Figure 8-3. The

6. Lockshon, D. and Morris, D.R. 1983. *Nucleic Acid Res.* **11:** 2999–3117.

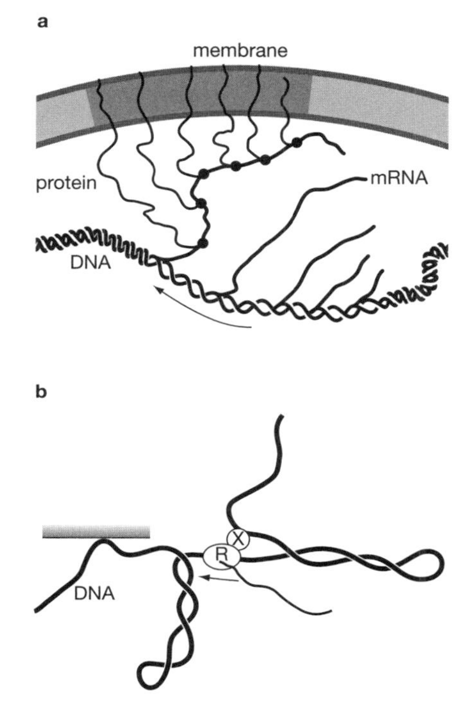

Figure 8-3. Some mechanisms that may contribute to transcriptional supercoiling of the DNA template. (*a*) Insertion of nascent polypeptides of a membrane protein (or a protein for export across the cell membrane) while the polypeptides are being extended along an elongating transcript (cotranscriptional membrane attachment or "transertion"). The arrow below the DNA indicates the direction of transcription, and multiple transcripts of the same gene have been initiated and elongated before an earlier transcript reaches its termination site. (Redrawn, with permission, from Binenbaum, Z., et al. 1999. *Mol. Microbiol.* **32:** 1173–1182, ©Blackwell Publishing Ltd.) (*b*) Association of the RNA polymerase with a DNA-bound protein X after the start of transcription. A negatively supercoiled loop forms behind the polymerase, and the same number of positive supercoils also accumulate in the region ahead of the polymerase if a downstream region of the DNA is anchored to a large cellular structure. (Redrawn, with permission, from Wang, J.C. 1987. *Harvey Lect.* **81:** 93–110.)

first is based on the coupling of transcription and translation in bacteria. Translation of a bacterial messenger RNA often starts before the transcribing RNA polymerase completes the RNA chain. In the case of a transcript encoding a membrane protein or a protein designated for export through the cell membrane, the signal peptide that targets the protein to the cell membrane is located at the amino terminus of the protein, and is thus synthesized at the very beginning of mRNA translation. Consequently, the amino-terminal region of a membrane protein or a protein designated for export is often inserted into the cell membrane long before mRNA synthesis is completed. Once this nascent polypeptide is inserted into the cell membrane, the transcription assembly R becomes anchored to the membrane, and the RNA polymerase tethered to the mRNA can no longer revolve around the DNA template (Fig. 8-3a). This cotranscriptional insertion of nascent polypeptides into cell membrane, which has been termed "transertion,"[7] appears to be one key mechanism that prevents the circling of R around DNA.[8]

7. Norris, V. 1995. *Mol. Microbiol.* **16:** 1051–1057.

8. Lodge, J.K., et al. 1989. *J. Bacteriol.* **171:** 2181–2187; Lynch, A.S. and Wang, J.C. 1993. *J. Bacteriol.* **175:** 1645–1655.

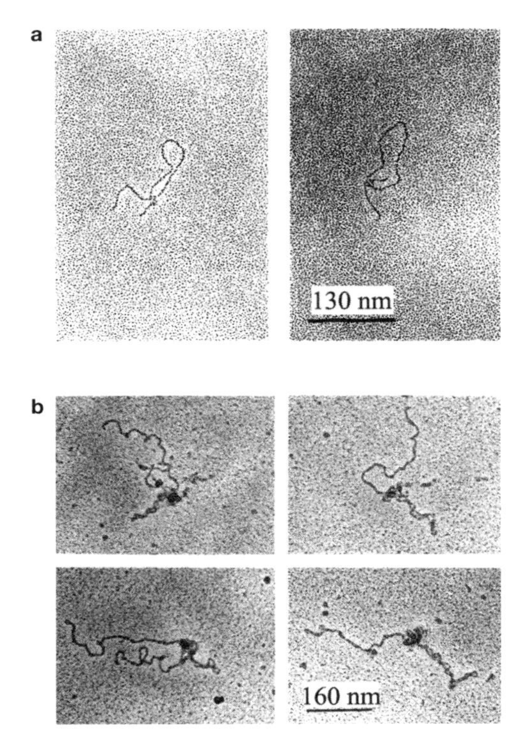

Figure 8-4. Interaction of a Gal4-T7 RNA polymerase fusion protein with its DNA binding sites. The DNA binding protein Gal4 was fused to T7 RNA polymerase and the resulting fusion protein incubated with a DNA fragment that carries a cluster of binding sites for Gal4 and a T7 promoter. See text for details. (*a*) The Gal4-T7 RNA polymerase fusion is incubated with the DNA fragment carrying Gal4 binding sites and T7 promoter. (*b*) The fusion protein is bound as in *a*, then incubated in a reaction mixture that promotes transcription from the T7 promoter. (Reprinted, with permission of the American Association for the Advancement of Science, from Ostrander, E.A., et al. 1990. *Science* **249:** 1261–1265.)

In the second mechanism, it is postulated that rotation of a transcribing RNA polymerase around its template is prevented by its interaction with a DNA-bound protein (Fig. 8-3b).[9] In the figure, the RNA polymerase is shown to interact with a protein X tightly bound to the same DNA, and this interaction encloses a DNA loop between the polymerase and X. Because of this anchoring of the polymerase to the DNA template itself, circling of the polymerase around the DNA is not possible; thus as transcription proceeds, the DNA is cranked through the polymerase associated with X, and the loop between the two proteins is lengthened and becomes negatively supercoiled at the same time. Furthermore, if the DNA ahead of the transcribing polymerase is also anchored, then positive supercoils would accumulate ahead of the advancing polymerase.

An experimental demonstration of this mechanism is shown by the electron micrographs depicted in Figure 8-4.[10] In this experiment, phage T7 RNA polymerase had been covalently joined to a yeast protein Gal4, a DNA binding protein that binds tightly to a specific DNA sequence. In each of the two electron micrographs shown

9. Wang, J.C. 1987. *Harvey Lect.* **81:** 93–110.

10. Ostrander, E.A., et al. 1990. *Science* **249:** 1261–1265.

in Figure 8-4a, a T7-Gal4 fusion protein was first bound to a 1000-bp long DNA fragment bearing three Gal4 binding sites clustered in a region ~250 bp from one end of the DNA fragment, and a phage T7 promoter ~120 bp from the other end. Binding of the fusion protein to the Gal4 and the T7 promoter sites led to the formation of a DNA loop, and in the micrographs the fusion protein appears as a dark shape at the base of the DNA loop. Figure 8-4b depicts four images of the DNA bound fusion protein after incubation with a mixture of four ribonucleoside triphosphates to initiate RNA synthesis; it can be seen that the DNA loop in these micrographs had apparently become tightly supercoiled. This type of mechanism has been suggested for the regulation of human c-*MYC* expression, in which the supercoiling of a loop, which is enclosed by interactions between a transcribing polymerase and two regulatory proteins that bind to an upstream region of the gene, is believed to modulate the level of c-*MYC* transcription.[11]

In eukaryotes, it has also been suggested that RNA polymerase molecules engaged in active transcription may be localized to large transcription "factories,"[12] and transcription may thus involve the threading of a DNA through such an immobilized factory, and supercoiled domains may be generated as a consequence.

Barriers to DNA Axial Rotation

In the illustration shown in Figure 8-2, the barriers R and E are shown to divide a DNA ring into two arcs of oppositely supercoiled regions. Inside an *E. coli* cell, it has been shown that for a DNA ring several thousand base pairs in length and expresses a single *tetA* gene and no other transcription units, hypernegative supercoiling of the ring in an *E. coli topA* mutant does not occur. Thus in the absence of E a connecting DNA segment several kilobases in length does not provide a friction barrier sufficiently large against the merge of two oppositely supercoiled domains. What kinds of elements might serve as the role of the hypothetical friction barrier E?

In the simplest case, E can be a second transcription assembly R′, which, like R, is also incapable of circling around the DNA template. One experiment used a construct based on a plasmid bearing a pair of inducible promoters that are closely spaced and transcribed in opposite directions.[13] If both promoters expressed messenger RNAs encoding membrane proteins, then, in a *topA* mutant lacking DNA topoisomerase I, a rapid accumulation of negative supercoils was seen in the plasmid on inducing the promoters. No such accumulation was evident, however, on inducing the same promoters that drive the synthesis of mRNAs that do not encode membrane proteins.[13] Because the rate of RNA chain elongation by *E. coli* RNA polymerase is about 40 nucleotides per second, a pair of RNA polymerases moving away from each other are

11. Liu, J., et al. 2006. *EMBO J.* **25:** 2119–2130.

12. Iborra, F.J., et al. 1996. *Exp. Cell Res.* **229:** 167–173.

13. Cook, D.N., et al. 1992. *Proc. Natl. Acad. Sci.* **89:** 10603–10607.

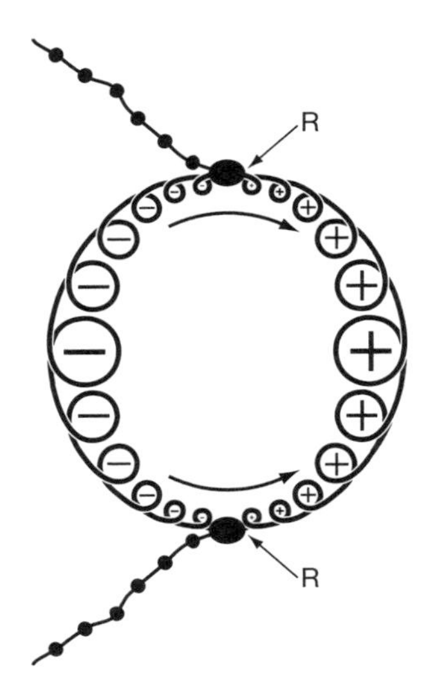

Figure 8-5. Transcription assemblies proceeding in opposite directions along a DNA ring. Here two transcriptional assemblies (R) are shown to move in opposite directions (indicated by the arrows) along a ring-shaped double-stranded DNA, represented by a single line. Each advancing R generates positive supercoils (+ signs) ahead of it and negative supercoils (– signs) behind it, which accumulate between the two assemblies.

expected to traverse a total of about 80 bp per second, and thus accumulate about eight negative supercoils per second behind the pair of transcription assemblies. The observed rate of accumulation of negative supercoils on induction of the promoters that drive the synthesis of mRNAs encoding membrane proteins is consistent with this expectation, suggesting that in the *topA* mutant the positive but not the negative supercoils are very efficiently removed by gyrase and Top4.[13] This supercoiling of a DNA ring by a pair of transcription assemblies requires that they move in opposite directions along the DNA ring (Fig. 8-5); otherwise, the positive supercoils generated in front of one assembly would cancel the negative supercoils in the wake of the other, with no accumulation of negative supercoils, even in a *topA* mutant.[5]

The friction barriers are not limited to transcription assemblies. In general, any DNA-bound protein that interacts with a membrane-embedded entity or any other large cellular structure can constitute a large barrier to axial rotation of the DNA. Furthermore, although a connecting DNA segment several kilobases in length does not appear to constitute a sufficient barrier to the cancellation of oppositely supercoiled domains, DNA bends or kinks that are stabilized by proteins may greatly increase the difficulty of rotating the DNA segment around its helical axis, and may thus serve the role of the element E depicted in Figure 8-2.[14] The anchoring of the RNA polymerase itself to its template, as described earlier, is also very effective in preventing the circling of the transcription assembly around the template DNA (Figs. 8-3b and Fig. 8-4).

14. Leng, F. and McMacken, R. 2002. *Proc. Natl. Acad. Sci.* **99:** 9139–9144; Stupina, V. and Wang, J.C. 2004. *Proc. Natl. Acad. Sci.* **101:** 8608–8613.

Association of DNA Topoisomerases with Actively Transcribed Regions

From the previous several sections, it is clear that transcription may drive DNA supercoiling under certain conditions, and, when that occurs, positive supercoils are generated ahead of a transcription assembly and negative supercoils are generated behind it. As we have seen, these oppositely supercoiled domains can be eliminated by their diffusional cancellation, or by the action of one or more DNA topoisomerases.

The generation of oppositely supercoiled domains by transcription does not necessarily predicate, however, that the DNA topoisomerases have a significant role in transcription. For example, even though an anchored R may lead to axial rotation of the DNA template and thus supercoiling of the template, such an anchor may periodically disappear whenever R dissociates from the DNA at the end of a round of transcription. In regions close to a chromosomal end, the supercoils may also dissipate by twirling the DNA if the end is not itself tightly anchored to the nuclear or cell membrane.

On the other hand, the DNA topoisomerases are known to associate with actively transcribed regions of intracellular DNA, and this association suggests that these enzymes may indeed play significant roles in transcription. In certain secretory cells of insects, such as the salivary glands of *Drosophila*, each diploid chromosome pair undergoes multiple rounds of replication to form a chromosome bundle that is visible by optical microscopy. In such a polytene chromosome, as many as a thousand copies of DNA molecules are laterally aligned and are sequence-wise in register. Actively transcribed regions of these large polytene chromosomes are less condensed, and can be seen under a light microscope as morphologically characteristic "puffs." By using antibodies specific to *Drosophila* Top1 to stain such polytene puffs, it has been shown that the enzyme is preferentially associated with the puffs.[15] This association is particularly striking at the heat shock loci, where transcription can be induced by a sudden elevation in temperature. In addition to RNA polymerase II, the enzyme responsible for mRNA synthesis in eukaryotes like the fruit fly, Top1 is also recruited to such heat-shock gene loci during but not before the heat-shock response: Staining of these loci by Top1-specific antibodies occurs after the heat-shock induction but not before. A higher resolution analysis of the locations of the topoisomerase molecules within a heat-shock locus shows that they are associated with the transcribed region, but not the nontranscribed flanking DNA sequences.[15] In human ribosomal RNA genes, which are transcribed by RNA polymerase I rather than RNA polymerase II, the locations of Top1 also show a very similar pattern, indicating that the enzyme is preferentially associated with regions transcribed by RNA polymerase I or II.[16] The localization of a *Drosophila* Top2-

15. Fleischmann, G., et al. 1984. *Proc. Natl. Acad. Sci.* **81:** 6958–6962; Gilmour, D.S., et al. 1986. *Cell* **44:** 401–407; Kroeger, P.E and. Rowe, T.C. 1992. *Biochemistry* **31:** 2492–2501.

16. Zhang, H., et al. 1988. *Proc. Natl. Acad. Sci.* **85:** 1060–1064.

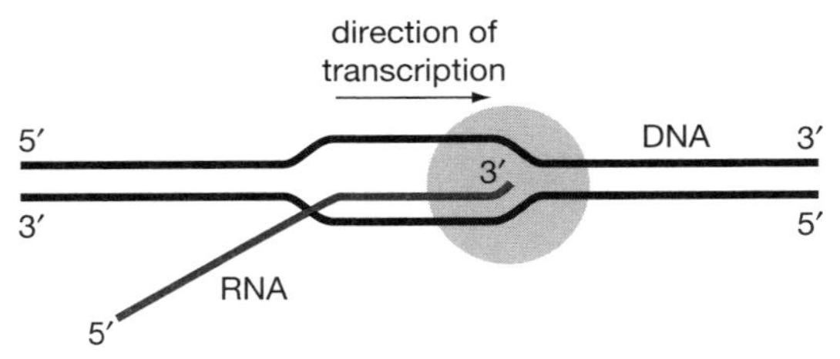

Figure 8-6. The formation of an "R-loop" structure during transcription. The pair of black lines represents a double-stranded DNA with a single-stranded bubble in it; for simplicity, the flanking double-helical regions are shown as parallel rather than intertwined lines. The shadowed sphere represents an RNA polymerase transcribing the DNA template in the direction indicated by the arrow, and the blue line represents the nascent RNA transcript. During transcription, the RNA polymerase adds ribonucleotides to the 3′ end of the RNA chain. To properly transcribe the nucleotide sequence of the template DNA strand, several RNA–DNA base pairs at the 3′ end of the nascent RNA are normally formed within a transcribing polymerase. Excessive negative supercoiling of the DNA behind the polymerase destabilizes the DNA double helical structure, however, and consequently a much longer RNA–DNA heteroduplex may form.

associated protein SCF to natural polytene puffs as well as puffs induced by a hormone ecdysteroid is again suggestive of a role of the topoisomerases in transcription.[17]

The association of a DNA topoisomerase with actively transcribed regions of intracellular DNA is further exemplified by studies implicating a physical association between a transcribing RNA polymerase and a type IB or IA DNA topoisomerase in *Drosophila* and *E. coli*, respectively.[18] In *E. coli*, the association involving the β′ subunit of the RNA polymerase and the carboxy-terminal domain of *E. coli* Top1 probably helps to position the type IA enzyme behind an advancing RNA polymerase, so as to remove the negative supercoils generated by the transcribing polymerase.

The presence of one or more DNA topoisomerases in regions undergoing transcription is presumably closely related to the generation of supercoils by transcription. If the supercoils are allowed to accumulate, cellular processes involving DNA, including the transcription process itself, are likely to be severely affected. Accumulation of positive supercoils in a yeast plasmid, for example, has been shown to abolish its transcription inside yeast cells.[19] In *E. coli*, there is evidence that excessive negative supercoiling of the region behind R may exacerbate what is termed "R-looping"—that is, the formation of an extended hybrid between the template strand of the DNA and the nascent RNA being elongated by the RNA polymerase (Fig. 8-6).[20]

17. Kobayashi, M., et al. 1998. *Mol. Cell Biol.* **18:** 6737–6744.

18. Shaiu, W.L. and Hsieh, T.S. 1998. *Mol. Cell Biol.* **18**: 4358–4367; Cheng, B., et al. 2003. *J. Biol. Chem.* **278:** 30705–30710.

19. Gartenberg, M. and Wang, J.C. 1992. *Proc. Natl. Acad. Sci.* **89:** 11461–11465.

20. Drolet, M., et al. 1994. *J. Biol. Chem.* **269:** 2068–2074.

The preferential association of the type IB enzyme Top1 with actively transcribed genes in *Drosophila* and in human cells suggests that this enzyme has a major role in removing the positive and negative supercoils generated by transcription in organisms in which it is expressed. Surprisingly, in yeast the deletion of the *TOP1* gene encoding the sole type IB enzyme of this organism has minimal effect on transcription. This dispensability can be attributed at least in part to the existence of diffusional pathways for the cancellation of positive and negative supercoils, and to the ability of Top2 to substitute for Top1: Inactivation of both yeast Top1 and Top2 has been shown to effect a large reduction in rRNA synthesis, as well as a significant reduction in mRNA synthesis.[21] The roles of the topoisomerases in gene expression are not limited to preventing excessive supercoiling of the DNA template, however, and some of these additional roles will be described in the next section and in Chapter 10.

DNA TOPOISOMERASES AND GENE EXPRESSION: BEYOND TRANSCRIPTIONAL SUPERCOILING

In yeasts and *Drosophila*, there is only one type II DNA topoisomerase, and its indispensable role in chromosome segregation overshadows any other physiological roles it might also have. In vertebrates, however, there are two closely related type IIA DNA topoisomerases, Top2α and Top2β, with very different expression patterns. A striking example of this difference can be seen during embryonic development of the mouse brain.

In the laminated structure of a developing mammalian cortex, neuronal precursor cells first undergo divisions in a layer called the ventricular zone. The postmitotic cells then migrate outward to the superficial layers, and in that process passing through the layer termed the intermediate zone as well as the cortical layers that have been formed by neurons born at an earlier time. Whereas antibodies specific to Top2α prominently stain the ventricular zone where the neuronal cells undergo proliferation, antibodies specific to its congener Top2β primarily stain layers containing the postmitotic neurons in the cortical layers but not the ventricular and intermediate zones.[22] Similarly, during postnatal development of the rat cerebellum, a sharp transition from Top2α to Top2β expression is seen in cell populations that have undergone the final cell division and are committed to differentiate into Purkinje cells, large neurons that form the Purkinje layer in the cerebellum, or granule cells, small neurons that are found in the cerebellum granule layer as well as several other regions of the brain.[23]

21. Uemura, T. and Yanagida, M. 1986. *EMBO J.* **5:** 1003–1010; Brill, S.J., et al. 1987. *Nature* **326:** 414–416; Brill, S.J. and Sternglanz, R. 1988. *Cell* **54:** 403–411.

22. Lyu, T.L. and Wang, J.C. 2003. *Proc. Natl. Acad. Sci.* **100:** 7123–7128.

23. Tsutsui, K., et al. 2001. *J. Comp. Neurol.* **431:** 228–239; Tsutsui, K., et al. 2001. *J. Biol. Chem.* **276:** 5769–5778; for a review, see Tsutsui, K.M., et al. *Anat. Sci. Int.* **81:** 156–163.

As we saw in Chapter 7, the close association of mammalian Top2α with proliferating cells suggests that this particular isozyme, similar to the single type II DNA topoisomerase in the yeasts, plays a critical role in mitosis. Indeed, studies in human cells show that during mitosis chromosome condensation and segregation are severely affected on inactivation of Top2α.[24] Given this crucial role of Top2α in cell division, it is not surprising that mouse *top2α*$^{-/-}$ embryos lacking the enzyme are found to perish before the eight-cell stage (see Footnote 7 of Chapter 7). But what then might be the roles of the IIβ isozyme, as its absence in proliferating cells clearly indicates that it is neither involved in DNA replication nor has a critical role in chromosome segregation?

Recall also from the discussion in Chapter 7 that *top2β*$^{-/-}$ mutant embryos develop to term but die at birth. Cell differentiation and organ development are largely normal in the mutant lacking the enzyme, and death of the newborns is apparently caused by defective neural development that leads to a breathing impairment. In the mutant embryos, motor axons fail to innervate limb and diaphragm muscles, and sensory axons fail to enter the spinal cord. Thus, during development Top2β seems to exert specific rather than general effects on genetic programming. Malformation of the cortical layers, owing to aberrant migration of the postmitotic neurons, is also observed in the mouse *top2β*$^{-/-}$ knockout mutant, or in mutant mouse embryos in which the *TOP2β* gene is deleted only in the developing brain.[22]

Because the laminar locations and the properties of the cortical neurons in the embryonic brain are mainly determined during the last cell division of the neuron progenitors[25] (before the elevation of Top2β expression level in those cells), it appears that Top2β does not have a direct role in laminar fate determination of the neurons. Rather, the enzyme may directly or indirectly affect the expression of a subset of genes when the postmitotic neurons further differentiate to become progressively more specialized.

The defective lamination of the cortical neurons in *top2β*$^{-/-}$ embryos is reminiscent of that caused by the consequence of inactivating the mouse *Reelin* gene. Expression of the Reelin mRNA is indeed much reduced in the *top2β*$^{-/-}$ brains, but Top2β inactivation apparently affects more than Reelin expression. Unlike the *Reelin* mutants that are viable, the brain-specific *top2β*$^{-/-}$ knockout mutants, which do not express Top2β in the brain but express the enzyme in all other tissues, nevertheless die at birth just like embryos lacking a functional *TOP2β* gene in all tissues. An examination of the gene expression profiles in embryonic brains of the *top2β*$^{-/-}$ mutants and their wild-type littermates shows that about 1%–4% of all expressed genes are either up- or down-regulated in the mutant brains at different developmental stages.[26] These affected genes encode proteins with diverse functions, including cell migration and adhesion, regulation

24. Grue, P., et al. 1998. *J. Biol. Chem.* **273:** 33660–33666.

25. McConnell, S.K. and Kaznowski, C.E. 1991. *Science* **254:** 282–285.

26. Lyu, Y.L., et al. *Mol. Cell Biol.* **26:** 7929–7941.

of transcription, ion transport, etc. No particular pattern in the affected genes is discernible in terms of either their physical or functional characteristics.

Although these findings specifically implicate an effect of Top2β on gene expression during brain development, it leaves open the possibility that Top2α and Top2β might substitute for each other in cells that express both. Whichever the case may be, the following question remains the same: Why should a type II DNA topoisomerase affect the expression of certain genes? Relieving template supercoiling by such an enzyme is often invoked, but this interpretation is unsatisfying. It does not account for the need for a type II enzyme in the presence of the type IB enzyme Top1, which is fully capable of removing both positive and negative supercoils. In vitro experiments have suggested that a type II enzyme might be more efficient than a type IB enzyme in relieving supercoils in a chromatin template decorated with nucleosomes.[27] But if a type II DNA topoisomerase is specifically required to relieve template supercoiling during chromatin transcription, why is RNA synthesis in the yeasts little affected by inactivation of DNA topoisomerase II?[21] And why is the effect of inactivating mouse Top2β on gene expression in postmitotic neurons limited to a few percent rather than the majority of the genes?[26]

But if a type II DNA topoisomerase affects gene expression beyond its role in supercoil removal, what might be the underlying mechanism? There are several possibilities to consider. First, because intracellular DNA exists as a complex and dynamic nucleoprotein entity, the expression of a gene is intricately related to its nucleoprotein architecture and dynamics. Both human Top2α and Top2β have been found to interact with HDAC1 and HDAC2, two proteins which participate in chromatin remodeling by removing the acetyl group from the nucleosome core histones.[28] Association between Top2β and a human chromatin remodeling complex hACF has also been reported.[29] Thus it is plausible that the Top2 isozymes, as components of larger complexes, may modulate interactions between various remodeling machineries and their chromatin targets, and consequently alter nucleosome positioning as well as interactions between the DNA template and various components of the transcription machinery. It may be significant that, in the developing embryonic brain, the enzymes HDAC2 and Top2β are both preferentially expressed in postmitotic neurons.[26]

Second, type II DNA topoisomerases have long been implicated in chromatin folding and organization (see the section "Role of DNA topoisomerases in chromosome condensation" in Chapter 7). The effects of the type II DNA topoisomerases on genetic programming may reflect intricate links among gene expression, chromosome folding, and chromosome condensation and decondensation. Changes in gene expression patterns during differentiation, for example, may require local or regional alterations of chromatin structures, and condensation or decondensation of chromosomal regions to

27. Mondal, N. and Parvin, J.D. 2001. *Nature* **413:** 435–438; Salceda, J., et al. 2006. *EMBO J.* **25:** 2575–2583.

28. Tsai, S.C., et al. 2000. *Nat. Genet.* **26:** 349-353; Johnson, C.A., et al. 2001. *J. Biol. Chem.* **276:** 4539–4542.

29. LeRoy, G., et al. 2000. *J. Biol. Chem.* **275:** 14787–14790.

different degrees of compaction may involve a type II DNA topoisomerase. A hint in this direction is the finding that *Drosophila* Top2 interacts with a condensin component *barren* involved in chromatin compaction: the Top2-*barren* pair may bind to particular sites and affect the expression of a group of genes that are involved in the regulation of the body plan of the fruit fly.[30]

At the present time both of these conjectures are rather vague, and the underlining molecular mechanisms remain largely unknown. Nevertheless, several possible explanations are suggested by the known characteristics of these enzymes. A type II DNA topoisomerase, for example, could act primarily as a DNA-binding component in larger complexes that target various other proteins to particular chromatin sites. In such a scenario, the enzyme's ability to move one DNA double helix through another might not be essential. An analogous example is provided by the finding that either human Top1, or an inactive mutant of it in which the active-site tyrosine is substituted by a phenylalanine, can augment the repression of the basal level transcription and the activation of transcription of a special class of promoters.[31] It has also been postulated in various contexts that a type II DNA topoisomerase, perhaps in association with other proteins, may act as a site-specific DNA supercoiling activity and thereby affect various actions at such sites.[32,33] Another interesting possibility is related to the accumulation of evidence that gene expression often involves formation and disruption of chromatin loops.[34] As discussed earlier, a type II DNA topoisomerase could function as a loop fastener, or using its DNA transport activity to facilitate loop formation.

It has also been suggested that Top2β may play a critical role in the activation of a number of regulated genes.[35] In the case of the estrogen inducible gene *p52*, for example, inhibiting the level of the DNA topoisomerase was shown to prevent the induction of the gene, which led to the suggestion that the Top2β-mediated formation of a transient double-stranded break in the promoter region might be required to initiate a chain of events beginning with the activation of poly-[ADP-ribose] polymerase-1.[36] Further experiments are needed, however, to test more critically the idea of gene induction through a topoisomerase-mediated transient break.

---

30. Bhat, M.A., et al. 2001. *Cell* **87:** 1103–1114; Lupo, R., et al. 2001. *Mol. Cell* **7:** 127–136.

31. Merino, A., et al. 1993. *Nature* **365:** 227–232.

32. Liu, L.F., et al. 1979. *Nature* **281:** 456–461.

33. Ohta, T. and Hirose, S. 1990. *Proc. Natl. Acad. Sci.* **87:** 5307–5311; Kobayashi, M., et al. *Mol. Cell Biol.* **18:** 6737–6744; Furuhashi, H., et al. 2006. *Development* **133:** 4475–4483.

34. Ptashne, M. and Gann, A. 2002. *Genes and signals*. Cold Spring Harbor Laboratory Press, Cold Spring Harbor, New York; Carter, D., et al. *Nat. Genet.* **32:** 623–626; Tolhuis, B., et al. 2002. *Mol. Cell* **10:** 1453–1465; Spilianakis, C.G., et al. 2995. *Nature* **435:** 637–645; Horike, S., et al. 2005. *Nat. Genet.* **37:** 31–40; Cai, S., et al. 2006. *Nat. Genet.* **38:** 1278–1288.

35. See, for example, Reitman, M. and Felsenfeld, G. 1990. *Mol. Cell Biol.* **10:** 2774–2786; Ju, B.G., et al. 2006. *Science* **312:** 1798–1802.

36. Ju, B.G., *et al.*, see Footnote 35.

CHAPTER 9

Beyond DNA Untanglement
New Twists on an Old Theme

"Nature does nothing useless."
Aristotle

As we saw in the previous chapters, the DNA topoisomerases were among the earliest pioneers of the DNA world, and their ascendancy was likely to be closely tied to the gradual lengthening of DNA and the emergence of ring-shaped DNA molecules. The type IA enzymes, and probably the type IB enzymes as well, were on the scene before the divergence of bacteria, eukarya, and archaea. As for the type II DNA topoisomerases, the last common ancestor of bacteria already had a type IIA enzyme in its tool chest, and the last common ancestor of archaea is likely to possess a type IIB enzyme.[1]

In the earlier years of the DNA topoisomerases, their mission was probably largely limited to alleviating DNA entanglement problems, which they accomplished splendidly. But their gradual ascendancy also opened new possibilities in the DNA world, as we shall see in this chapter. Nature, ever eager to explore new possibilities that arise from its countless trials of twiddling and tweaking its inventions, would hardly be expected to resist new opportunities in one of its most treasured gems—the material of which genes are made.

In the preceding chapters, our emphasis has been on how Nature responded to problems manifested by the coiling and entanglement of intracellular DNA. Here we shall shift our focus to how Nature might have taken advantage of some of these man-

1. Reviewed in Brochier-Armanet, C., et al. 2008. *Nat. Rev. Microbiol.* **6:** 245–252; Gadelle, D., et al. 2003. *BioEssays* **25:** 232–242; Forterre, P., et al. 2007. *Biochimie* **89:** 427–446. See also Brochier-Armanet, C. and Forterre, P. 2007. *Archaea* **2:** 83–93.

ifestations to turn a liability into an asset. In this respect, DNA supercoiling is by far the most important and the best-studied example.

DNA SUPERCOILING

A Much More Lively DNA in Its Negatively Supercoiled State

As we saw in the preceding chapter, DNA inside a bacterium like *Escherichia coli* is negatively supercoiled. Waves of positive supercoils, driven by replication, transcription, and perhaps other processes involving a macromolecular assembly tracking along the helical DNA, may come and go at various locations and at various times. There is strong evidence, however, that intracellular bacterial DNA is negatively supercoiled, when averaged over all locations or when averaged over time in a particular region. Similarly, intracellular DNA in archaea expressing gyrase (or both gyrase and the type IA enzyme reverse gyrase) also appears to be negatively supercoiled at the physiological temperatures for these organisms.[2] In contrast to the negatively supercoiled state of DNA in prokaryotes expressing gyrase (bacteria and archaea), the bulk chromatin in eukaryotes, as well as DNA in archaea expressing reverse gyrase (but not gyrase), is generally found in a relaxed state.[2,3] Even in the latter cases, however, specific regions may exist in a supercoiled state. Examination of a number of regions undergoing active transcription indicates that supercoiling of intracellular DNA is not restricted to prokaryotes expressing gyrase.[4]

In the environment experienced by organisms that live at moderate temperatures (the "mesophiles"), the two intertwined strands in a DNA double helix are stably paired through hydrogen bonding between the bases attached to their backbones. It has been known for more than three decades, however, that this normally placid double-helix structure would become much more lively upon negative supercoiling. The strands of a DNA double helix could come apart more readily; a variety of unusual structures, including the left-handed Z-helical form and a three-stranded H-form, could pop into existence, and the DNA would be much more receptive to proteins that unwind the double helix when bound to it.[5] Positive supercoiling, on

2. Charbonnier, F. and Forterre, P. 1994. *J. Bacteriol.* **176:** 1251–1259: Forterre, P., et al. 1996. *FEMS Microbiol. Rev.* **18:** 237–248; López-García, P. 1999. *J. Mol. Evol.* **49:** 439–452; Guipaud, O., et al. 1997. *Proc. Natl. Acad. Sci.* **94:** 10606–10611.

3. Sinden, R.R., et al. 1980. *Cell* **21:** 773–783. The "relaxed state" refers to the DNA associated with bound cellular proteins, including the histones.

4. See, for example, Leonard, M.W. and Patient, R.K. 1991. *Mol. Cell. Biol.* **11:** 6128–6138; Ljungman, M. and Hanawalt, P.C. 1992. *Proc. Natl. Acad. Sci.* **89:** 6055–6059; Jupe, E.R., et al. 1993. *EMBO J.* **12:** 1067–1075; Kramer, P.R. and Sinden, R.R. 1997. *Biochemistry* **36:** 3151–3158; Liu, J., et al. 2006. *EMBO J.* **25:** 2119–2130.

5. See, for example, Vinograd, J., et al. 1968. *J. Mol. Biol.* **33:** 173–197; Bauer, W. and Vinograd, J. 1970. *J. Mol. Biol.* **47:** 419–435; Davidson, N. 1972. *J. Mol. Biol.* **66:** 307–309; Hsieh, T.S. and Wang, J.C. 1975. *Biochemistry* **14:** 527–535; Vologodskii, A.V. and Cozzarelli, N.R. 1994. *Annu. Rev. Biophys. Biomol. Struct.* **23:** 609–643; van Holde, K. and Zlatanova, J. 1994. *BioEssays* **16:** 59–68; Kouzine, F. and Levens, D. 2007. *Front. Biosci.* **12:** 4409–4423; Wells, R.D. 2007. *Trends Biochem. Sci.* **32:** 271–278.

the other hand, tends to stabilize the canonical DNA double-helix structure, and the effects of positive supercoiling on DNA–protein interactions are contrary to those of negative supercoiling.[5]

There is strong evidence that Nature has made ample use of these and other effects of supercoiling, and we shall consider several specific examples in later sections of this chapter. A better understanding of the biological effects of DNA supercoiling may be helped, however, by first reviewing a few basic properties of supercoiled DNA.

Some Basic Properties of Supercoiled DNA

A DNA ring provides a good example for illustrating some consequences of supercoiling, but the situation with linear chromosomes is not that different, owing to their great length and their organization into structures with many large loops. As described in Chapter 2, a supercoiled DNA ring is characterized by a parameter termed the linking number (Lk). The deviation of Lk from Lk^0, the latter being the linking number of the same DNA ring in a completely relaxed reference state, provides a metric for the extent, or degree, of positive or negative supercoiling. By definition, positive supercoiling refers to $Lk > Lk^0$ or $(Lk - Lk^0) > 0$, and negative supercoiling $Lk < Lk^0$ or $(Lk - Lk^0) < 0$.

Some consequences of supercoiling are probably intuitively apparent. As in the case of a rope with two twisted strands, a DNA ring becomes increasingly contorted as Lk deviates more and more from Lk^0. The contorted ring assumes a more compact form in space relative to the same molecule in its relaxed state, and hence its movements in a liquid medium encounter a lower frictional drag. Thus, upon supercoiling, a DNA ring would sediment faster in a centrifugal field, diffuse faster in a solution, and migrate more rapidly through a gel matrix. Some of these properties played a major role in the early studies of DNA topoisomerases. Measurements of the sedimentation velocities of a DNA supercoiled to different extents, for example, were instrumental in the discovery of the first member of this enzyme family (see Chapter 2), and differences in the electrophoretic mobility of DNA rings of different linking numbers in an agarose gel (see Fig. 5-2) were a key to establishing the basic mechanism of the type II DNA topoisomerases (see Chapter 5).

As the DNA double helix in a supercoiled DNA coils up to form ever tighter loops in which the DNA double helix doubles on itself, threading another linear DNA through the supercoiled DNA becomes more and more difficult. Thus, if a linear DNA is made to circularize in the presence of a DNA ring, the chance of forming a catenane would decrease drastically if the latter is in a tightly supercoiled form. Following the same logic, relative to their relaxed counterparts it would be much more probable for two supercoiled DNA rings to be unlinked rather than catenated.[6]

6. Rybenkov, V.V., et al. 1997. *J. Mol. Biol.* **267:** 312–323.

Other consequences of supercoiling are conceptually more abstract, and thus understanding these effects may require further consideration. Basically, a contorted supercoiled DNA is in a strained or higher energy state, like that of a tightly wound or compressed spring. A supercoiled DNA would have passed happily into a less contorted form but for the topological constraint that *Lk* be a constant. Thus, any process that helps to reduce the degree of supercoiling would occur more readily in a supercoiled DNA than in the same DNA without the topological constraints—for example, when the DNA ring is converted to a linear or nicked form, or an appropriate DNA topoisomerase is present to transiently break the DNA strands.

But how can the degree of supercoiling be changed if *Lk* is fixed for a DNA ring with unbroken strands? Because the degree of supercoiling of a DNA ring is determined by the deviation of *Lk* from Lk^0, and only *Lk* (and not Lk^0) is topologically fixed for a DNA ring with intact strands, the degree of supercoiling can be readily changed by altering Lk^0. For example, we saw in Chapter 2 that Lk^0 of a DNA would be reduced if the helical structure of the DNA were changed from the B-form (having 10 bp per turn) to the A-form (11 bp per turn), through the disruption of base pairing between the complementary strands, or by binding of an intercalator like ethidium, which inserts itself between two neighboring base pairs in a DNA double helix and unwinds the DNA by 26° (see Chapter 3). Lk^0 would be reduced by 1 if a DNA segment 110 bp long is changed from the normal B-form (110/10 or 11 turns), to the A-form (110/11 or 10 turns), if 10 bp of B-DNA (one helical turn) are disrupted, or if 14 ethidium molecules are inserted into the double helix ($26° \times 14 \approx 360°$ or one turn).

Qualitatively then, any process that changes Lk^0 in a way that makes a supercoiled DNA less supercoiled, and thus less contorted, would occur more readily in a supercoiled DNA than in a linear or nicked DNA. Thus, a process that reduces Lk^0 would be favored in a negatively supercoiled DNA, and a process that increases Lk^0 would be favored in a positively supercoiled DNA; in both cases, the magnitude of the linking difference $|Lk - Lk^0|$ is reduced by the stated change in Lk^0.

How large is the effect of supercoiling on such a process? Here the answer depends on how supercoiled the DNA is. It is intuitively clear that for the same numerical value of $|Lk - Lk^0|$, the strain a DNA ring experiences would depend on the size of the ring and hence the magnitude of Lk^0: changing *Lk* by 1 for a 10,000-bp DNA ring (with an Lk^0 of ~1000) would strain the ring only slightly, but the same change in *Lk* for a 200-bp ring (with an Lk^0 of ~20) would cause substantial strain in the molecule. Thus, the size-normalized quantity $\sigma \equiv (Lk - Lk^0)/Lk^0$, which is termed the specific linking difference, provides a better parameter for comparing the extents of supercoiling of DNA rings of different sizes. The values of σ for purified DNA rings isolated from *E. coli* are typically in the range –0.05 to –0.08; that is, their linking numbers are typically 5%–8% lower than we would expect for relaxed DNA rings of the same sizes.

For a DNA with a linking-number deficit in this range, reasonable estimates can be made of the effects of supercoiling on interactions between the DNA and other molecules, large or small, on the basis of experimental results that are omitted here.

A useful equation is $\log(r) \approx 91\sigma\Delta Lk^0$, where r is the relative binding constant of a molecule to a supercoiled DNA with a specific linking-number difference σ (the binding constant to the relaxed DNA is set to 1), and ΔLk^0 is the change in Lk^0 owing to the binding of this molecule. Note that $\log(r)$ is proportional to both σ and ΔLk^0, and thus the relative binding constant r is an exponential function of the product of σ and ΔLk^0.[7]

As an example, consider the binding of ethidium to a negatively supercoiled DNA with a σ of –0.06. The binding of an ethidium would reduce Lk^0 by 26°/360° or 0.072 turns (thus ΔLk^0 is –0.072); $\log(r)$ is therefore estimated to be 91(–0.06)(–0.072) or 0.33, which corresponds to an r of 2.5. Even though the binding of an ethidium molecule unwinds the DNA double helix by less than one-tenth of a turn, its binding constant to a typical negatively supercoiled DNA ring is increased by more than twofold![8] If a macromolecule unwinds a DNA by one full turn, the exponential dependence of r on ΔLk^0 would predict an enhancement in binding affinity to the same supercoiled DNA by nearly 1 million-fold![7]

Facilitating the Initiation of Replication by Negative Supercoiling

There is substantial evidence that a negatively supercoiled state of the DNA in a bacterium like *E. coli* is often necessary for initiation of replication. With the exception of replication that starts from an end (or ends) of a linear DNA, a key step in initiating replication in a double-stranded DNA involves the formation of a small replication "bubble," in which base pairing between a short segment of the two complementary strands is disrupted (see Fig. 7-4 in Chapter 7). Nature has invoked quite a few strategies to promote the formation of a replication bubble. At the *E. coli* replication origin *oriC*, for example, the strategy used is the binding of a few dozen molecules of a protein called DnaA.[9] For an *E. coli* plasmid called ColE1, a particular nascent transcript forms a short stretch of stable DNA–RNA "heteroduplex" with the template DNA strand at the repli-

7. Quantitative estimates of the effects of supercoiling in various processes are based on the energetics of supercoiling. The key parameter here is the "free energy change" of supercoiling. Energy is never free, and chemists use the term "free energy change" to specify the amount of energy from a chemical process that can be freely converted to useful work. Interested readers may consult introductory chemistry texts or, for more advanced discussions, thermodynamics treatises. For a primer on the free energy of DNA supercoiling and how it leads to the simple relation $\log r = 91\sigma\Delta Lk^0$, see Wang, J.C. 1994. *Adv. Pharmacol.* **29B:** 257–270.

8. This 2.5-fold higher affinity for the negatively supercoiled DNA, with a σ of –0.06, refers to the situation that arises when very few ethidium molecules are bound to each DNA ring. If increasing numbers of ethidium molecules were bound to the DNA, Lk^0 would progressively decrease. Correspondingly, $(Lk - Lk^0)$ would first become less and less negative, and then more and more positive after passing the point $(Lk - Lk^0) = 0$. The relative binding constant r at any point can be estimated by first calculating the value of σ from the calculated Lk^0 at that point, and then using the same equation $\log(r) \approx 91(\sigma)(-0.072)$. Qualitatively, as increasing numbers of ethidium molecules became bound to the negatively supercoiled DNA, the negatively supercoiled DNA would gradually relax and then become increasingly positively supercoiled; r would gradually decrease from 2.5 to 1, and would then become smaller than 1.

9. Crooke, E., et al. 1991. *Res. Microbiol.* **142:** 127–130.

cation origin of the plasmid, displacing the nontemplate strand and exposing it in a single-stranded form.[10] In both examples, negative supercoiling of the DNA destabilizes base pairing to facilitate the opening of the DNA duplex by DnaA, or displacement of the nontemplate strand by RNA. As we saw in Chapter 2, the formation of a single-stranded bubble in a DNA ring is helped by negative supercoiling of the DNA ($|Lk - Lk^0|$ is reduced). The use of negative supercoiling to open up an origin of replication is also advantageous in that negative supercoiling specifically destabilizes the parental double helix but has no effect on base pairing between the newly synthesized strand and its template, which is not topologically constrained owing to the presence of single-stranded regions in the newly replicated DNA.[11]

Whereas gyrase would be a logical choice for keeping the origin of replication in a negatively supercoiled state, alternative mechanisms are also plausible. For example, even before the discovery of gyrase, it was observed that initiation of phage λ replication requires transcription near its replication origin. Furthermore, this "transcriptional activation" depends on the transcription process itself rather than on any resulting protein product, and transcription can begin at least 95 bp downstream from *ori*λ, the replication origin of λ.[12] In light of the twin-supercoiled model of transcription (discussed in the preceding chapter), it seems likely that negative supercoiling of the DNA template in the wake of a transcription assembly may have a role in activating the phage λ replication origin.

In eukaryotes, intracellular DNA is generally thought to be in a relaxed form, but the possibility of local supercoiling suggests that, in this case also, initiation of replication may involve negative supercoiling.[3,4] The *Drosophila* origin recognition complex (ORC), for example, binds to a negatively supercoiled DNA by a couple of orders of magnitude more strongly than to a relaxed or linear DNA, and it has been suggested that localized supercoiling through remodeling of the nucleoprotein structure at the replication origin may have a significant role in the initiation of replication.[13]

Stabilization of DNA Base Pairing in a Hyperthermophile by Positive Supercoiling?

If negative supercoiling can be utilized to drive the unwinding of the DNA double helix, it would seem that positive supercoiling might be used to do the opposite—that is, to make the DNA double helix more stable and thus prevent its uncoiling. In

10. Itoh, T. and Tomizawa, J. 1979. *Cold Spring Harb. Symp. Quant. Biol.* **43:** 409–417.

11. Wang, J.C. 1978. Some aspects of DNA strand separation. In *DNA synthesis, present and future* (ed. I. Molineux and M. Kohiyama), pp. 347–366. Plenum Press, New York; Postow, L., et al. 2001. *Proc. Natl. Acad. Sci.* **98:** 8219–8226.

12. Furth, M.E., et al. 1972. *J. Mol. Biol.* **154:** 65–83; see also Leng, F. and McMacken, R. 2002. *Proc. Natl. Acad. Sci.* **99:** 9139–9144.

13. Remus, D., et al. 2004. *EMBO J.* **23:** 897–907.

hyperthermophiles that live at temperatures greater than 80°C, it had been suggested that a robust reverse gyrase activity might keep the intracellular DNA in a positively supercoiled state and thereby improve the stability of the base-paired state.[14]

Although DNA in many hyperthermophiles is usually found in a relaxed rather than a positively supercoiled state,[1] in terms of base pairing even a relaxed DNA ring (or loop) with intact strands is more stable than the same DNA in a nicked or linear state, because unpairing of the complementary strands in the former would reduce its Lk^0, and thus increase $|Lk - Lk^0|$, leading to positive supercoiling of the ring or loop.[15] Furthermore, the ATP-dependent type IA enzyme in the hyperthermophiles (reverse gyrase) may also act as an active DNA strand "renaturase"—that is, an enzyme that makes use of ATP to form right-handed braids between two single strands of DNA.[16] The strongest evidence for the renaturase idea comes from experiments using a DNA ring containing a small single-stranded "bubble," within which the two DNA strands are noncomplementary in their nucleotide sequences and therefore cannot form base pairs. These experiments show that a reverse gyrase can efficiently catalyze the ATP-dependent positive supercoiling of a DNA ring with a single-stranded bubble if the lengths of the pair of single strands are either 20 or 50 nucleotides, but not if the strands are only 5 nucleotides long.[16] Thus the enzyme reverse gyrase may recognize unpaired regions in intracellular DNA and rewind the separated strands, whether the DNA is in a relaxed state in a hyperthermophile or in a negatively supercoiled state in a gyrase-expressing mesophile.[16]

Positive Supercoiling, Replication-Fork Regression, and Lesion Bypass

An interesting possibility for utilizing positive supercoiling involves the retraction of a replication fork when it encounters a damaged site in the unreplicated DNA ahead of the fork (Fig. 9-1). A model first postulated in 1976[17] suggested that when a replication apparatus encounters a lesion on one template strand, copying of that strand would stop, whereas synthesis along the other strand can continue for a distance past the position of the lesion in the opposing strand (Fig. 9-1a). When the replication apparatus disassembles at the fork, the two newly synthesized progeny strands (which are complementary to one another because they are copied from complementary parental strands) can pair and grow in length as the replication fork retracts (Fig. 9-1b). This pairing could then allow the extension of the shorter progeny strand, using the

14. Kikuchi, A. and Asai, K. 1984. *Nature* **309:** 677–681.

15. Vinograd, J., et al. 1968. *J. Mol. Biol.* **33:** 173–197; Hsieh, T.-S. and Wang, J.C. 1975. *Biochemistry* **14:** 527–535.

16. Hsieh, T.S. and Plank, J.L. 2006. *J. Biol. Chem.* **281:** 5640–5647.

17. Higgins, N.P., et al. 1976. *J. Mol. Biol.* **101:** 417–425.

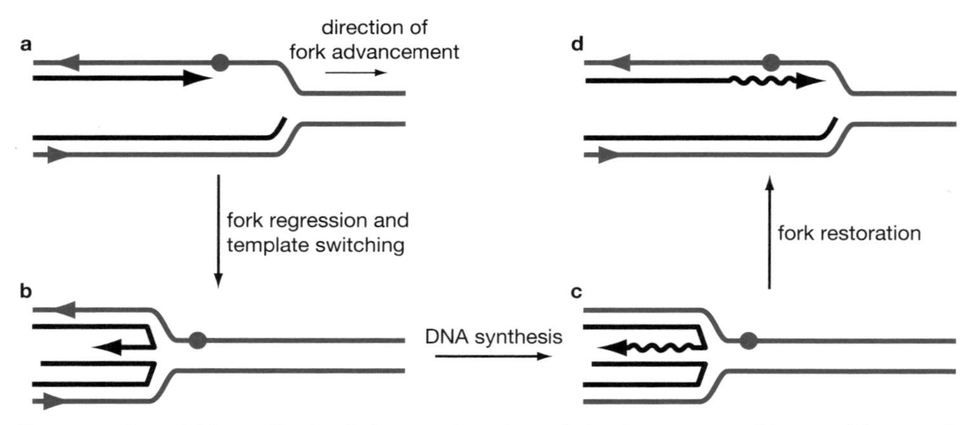

Figure 9-1. A model for replication fork regression when a lesion is encountered in one of the strands of a DNA double helix.[17] The blue lines represent the parental and the black lines the progeny DNA strands. For convenience, the complementary strands in the DNA double helix are represented by two parallel lines rather than by intertwined lines. The arrowhead in a line representing a DNA strand specifies the $5' \rightarrow 3'$ direction of the strand (it is known that DNA synthesis always proceeds in the $5' \rightarrow 3'$ direction by extension of a 3′-hydroxyl terminus). (*a*) Synthesis of one strand is blocked when the replication apparatus encounters a lesion (represented by the small blue circle) in one parental strand, but synthesis of the other strand proceeds past the position corresponding to the lesion in the opposite strand before the collapse of the replication fork. (*b*) The regression of the fork, favored by positive supercoiling of the unreplicated parental DNA ahead of the fork, and by base pairing between the newly synthesized progeny strands, leads to the formation of a four-way junction (a Holliday junction structure with one short arm). (*c*) Formation of the four-way junction provides a new template for extension of the shorter progeny strand (wavy line). (*d*) DNA synthesis can now switch back to copying the parental strands, bypassing the lesion in the damaged parental strand and leaving it for subsequent repair. (Redrawn, with permission of Elsevier, from Higgins, N.P., et al. 1976. *J. Mol. Biol.* **101:** 417–425.)

longer progeny strand as the template (Fig. 9-1c). Subsequently, the extended strand could be used to pair with the parental strand containing the lesion, thus bypassing the lesion for continued progression of the replication fork (Fig. 9-1d).[17] In this scheme, positive supercoiling of the DNA ahead of the replication fork would be expected to favor the formation of the four-way junction, because the retraction of the fork would relieve the overwinding of the parental strands (see the review by Postow et al. cited in Footnote 11). Because the four-way junction (sometimes dubbed the "chicken-foot") is structurally the same as a Holliday junction, the restoration of a functional replication fork may involve further processing of this structure by recombination and repair enzymes.[18]

18. See, for example, Seigeur, M., et al. 1998. *Cell* **95:** 419–430; Ralf, C., et al. 2006. *J. Biol. Chem.* **281:** 22839–22846.

Supercoiling and Chromosome Segregation

The compaction and condensation of intracellular DNA, and the segregation of chromosome pairs during cell division, have been discussed in the preceding chapter. In bacteria such as *E. coli*, two lines of evidence suggest that DNA condensation/compaction, chromosome segregation, and supercoiling are closely interrelated. The first derives from studies of the segregation defect of a number of *E. coli muk* gene mutants.[19] A complex formed by the three *muk* gene products, MukB, MukE, and MukF folds and compacts DNA into a shape stabilized by negative supercoiling of the DNA.[20,21] Interestingly, it has been found that the segregation defect of several *E. coli mukB* mutants can be compensated by mutations in the *topA* gene that increase the overall degree of negative supercoiling of intracellular DNA, suggesting that chromosome segregation is facilitated by DNA supercoiling.[22] The second line of evidence comes from the findings that a number of *muk* mutants are hypersensitive to the gyrase inhibitor novobiocin,[23] and that suppression of the segregation defect of *muk* mutants by mutations in *topA* can be reversed by reducing the cellular activity of gyrase with coumermycin, an analog of novobiocin.[23] These findings again support the existence of a link between bacterial chromosome segregation and DNA supercoiling.

Three plausible mechanisms may explain how chromosome segregation and DNA supercoiling are related. First, as discussed at the beginning of this chapter, supercoiling of a pair of linked DNA rings may directly facilitate their segregation, because supercoiling of the rings greatly decreases the probability of catenation of the rings and increases the probability of their decatenation.[6] Negative supercoiling favors the folding of the DNA into a shape normally assumed by the DNA in its complex with MukB–MukE–MukF. Therefore, in the absence of *E. coli* Top1, the normal role of the MukB–MukE–MukF complex may be less critical.[20] In other words, the effect of DNA supercoiling on chromosome segregation may be indirect, and the common denominator between chromosome supercoiling and segregation may be the compaction of the chromosome. Third, negative supercoiling of the DNA rings could stimulate their decatenation by bacterial DNA topoisomerase IV, and thus facilitate chromosome segregation (see the discussion in Chapter 6 on the removal of L- and R-braids). These possibilities are not mutually exclusive, and additional studies are needed to better answer the question of how supercoiling helps chromosome segregation in bacteria.

---

19. Niki, H., et al. 1992. *EMBO J.* **11:** 5101–5109; Yamanaka, K., et al. 1996. *Mol. Gen. Genet.* **250:** 241–251.
20. Petrushenko, Z.M., et al. 2006. *J. Biol. Chem.* **281:** 4606–4615.
21. The MukB–MukE–MukF protein complex is similar to the eukaryotic condensin proteins for reviews on the condensins and related proteins, see Nasmyth, K. and Haering, C.H. 2005. *Annu. Rev. Biochem.* **74:** 595–648; Hirano, T. 2006. *Nat. Rev. Mol. Cell. Biol.* **7:** 311–322.
22. Sawitzke, J.A. and Austin, S. 2000. *Proc. Natl. Acad. Sci.* **97:** 1671–1676.
23. Onogi, T., et al. 2000. *J. Bacteriol.* **182:** 5898–5901.

Effects of DNA Supercoiling on Transcription

As described earlier, the supercoiling of a DNA profoundly affects its interaction with many other molecules large or small. Many interactions in the transcription process, including the formation of various promoter–RNA polymerase complexes, the binding of regulatory factors to a promoter, the clearance of a polymerase from a promoter after its initiation of a transcript, and the termination of the transcript are thus potentially dependent on DNA supercoiling, probably to different extents and in some cases in opposite ways. Furthermore, the helical structures of different promoters and regulatory elements may also respond differently to changes in the level of supercoiling. Because of the multiple possibilities, it is difficult to quantitatively predict how DNA supercoiling might affect the transcription of a particular gene and to what extent.

It is well documented, however, that in bacteria like *E. coli* and *Salmonella typhimurium*, transcription is strongly modulated by the extent of supercoiling of intracellular DNA. An examination of the transcription of the entire *E. coli* genome indicates that a general decrease in the extent of negative supercoiling, resulting from inhibitiion of DNA gyrase, significantly decreases the expression of ~5% of the genes and increases the expression of ~2.5% of the genes.[24] In addition, the twin-supercoiled-domain model of transcription stipulates that the local degree of supercoiling may vary greatly along a chromosome, and may also vary at different times at a given chromosomal location (as we saw in the preceding chapter). For example, as shown by studies conducted since the late 1980s, the expression of a divergent pair of adjacent promoters can be coupled, and a normally silent gene can be activated by negative supercoiling of the DNA behind the nascent transcript of a neighboring gene.[25]

Because of the strong effects of DNA supercoiling on transcription, it is plausible that Nature may also use supercoiling in the coordinated control of groups of genes, particularly in prokaryotes expressing gyrase. The regulation of DNA supercoiling has been implicated, for example, to play a major role in adaptation to environmental stresses, such as changes in temperature, pH, and nutrients, as well as oxidative and osmotic stresses and other physical and chemical assaults.[26] Most studies of this involve measurements of the state of supercoiling during stress adap-

24. Peter, B.J., et al. 2004. *Genome Biol.* **5:** R87.

25. For reviews, see Lilley, D.M. and Higgins, C.F. 1991. *Mol. Microbiol.* **5:** 779–783; Lilley, D.M., et al. 1996. *Q. Rev. Biophys.* **29:** 203–225; Wu, H.Y. and Fang, M. 2003. *Prog. Nucleic Acid Res. Mol. Biol.* **73:** 43–68.

26. For reviews, see Rui, S. and Tse-Dinh, Y.C. 2003. *Front. Biosci.* **8:** d256–d263; Dorman, C.J. 2006. *Sci. Prog.* **89:** 151–166. See also López-García, P. and Forterre, P. 2000. *BioEssays* **22:** 738–746; Cheung, K.J., et al. 2003. *Genome Res.* **13:** 206–215; Travers, A. and Muskhelishvili, G. 2005. *Nat. Rev. Microbiol.* **3:** 157–169.

tation, as well as analyses of the consequences of modulating the cellular levels of proteins that affect supercoiling, including DNA gyrase, DNA topoisomerase I, and architectural proteins that modulate DNA conformation. Experiments in which *E. coli* cells are propagated for many generations over a long period also indicate that modulation of DNA topology may facilitate the optimization of a gene expression pattern in response to environmental changes.[27]

In eukaryotes, the action of the type IB enzyme keeps intracellular DNA largely in a relaxed state,[3] and evidence for Nature's use of DNA supercoiling in gene regulation is not as strong as that in bacteria. It has been suggested, however, that transcription in both bacteria and eukaryotes shares the common features of DNA unwinding at the initiation site as well as a change in the packaging of intracellular DNA by proteins, and DNA topology may therefore also be utilized in the fine tuning of eukaryotic transcription.[28] Furthermore, the general absence of gyrase or reverse gyrase does not preclude the possibility of localized supercoiling mediated by transcription (see the preceding chapter and Footnote 4), by other processes involving the tracking of macromolecular assemblies along DNA, or by altering the nucleoproteins in the vicinity of a promoter.[29] For example, it has been shown that in a frog oocyte, transcription of a ribosomal RNA gene on a linear DNA template can be activated by a transcript that moves in the opposite direction, as in the coupling of adjacent gene expression in bacteria.[30] The idea of a type II DNA topoisomerase promoting supercoiling at specific locations can be traced back to 1979,[31] and it has been reported that *Drosophila* Top2, in the presence of a supercoiling factor (SCF), can negatively supercoil DNA and activate the transcription of particular genes.[32]

Another interesting question is whether supercoiling may facilitate the transcription of a eukaryotic DNA decorated with nucleosomes. The formation of a nucleosome is associated with a decrease of Lk^0 of $\sim$1; hence, nucleosome formation is greatly facilitated by negative supercoiling, and its disruption is greatly facilitated by positive supercoiling. In vitro experiments suggest that as an RNA polymerase advances, positive supercoiling of the DNA ahead of it might help to dislodge the nucleosome ahead of it, and negative supercoiling of the DNA be-

27. Wright, B.E. 2004. *Mol. Microbiol.* **52:** 64–650; Crozat, E., et al. 2005. *Genetics* **169:** 523–532; Crozat, E., et al. 2007. *BioEssays* **29:** 846–860.

28. Travers, A. and Muskhelishvili, G. 2007. *EMBO Rep.* **8:** 147–151, 2007. See also Liu, J., et al. 2006. *EMBO J.* **25:** 2219–2130.

29. Reviewed in Esposito, F. and Sinden, R.R. 1988. *Oxf. Surv. Eukaryot. Genes* **5:** 1–50; Freeman, L.A. and Garrard, W.T. 1992. *Crit. Rev. Eukaryot. Gene Expr.* **2:** 165–209.

30. Dunaway, M. and Ostrander, E.A. 1993. *Nature* **361:** 746–748.

31. Liu, L.F., et al. 1979. *Nature* **281:** 456–461.

32. Ohta, T. and Hirose, S. 1990. *Proc. Natl. Acad. Sci.* **87:** 3375–4483; Furuhashi, H., et al. 2006. *Development* **133:** 4475–4483; Ogasawara, Y., et al. 2007. *Genes Cells* **12:** 1347–1355.

hind it might help direct the transfer of the dislodged nucleosome to a position behind the polymerase.[33]

DNA Topology and Site-Specific Recombination

Historically, it was the observation that DNA must be negatively supercoiled in the phage λ integrase-mediated recombination between two unique sites on the same DNA that led to the discovery of DNA gyrase (see Chapter 5). Site-specific recombination usually involves the formation of a DNA–protein complex, in which a recombinase, often augmented by other protein molecules, interacts with a pair of specific sequences to mediate the exchange of DNA strands. In the case of the λ integrase-mediated reaction, the two sites are termed *att*P, the phage attachment site normally present on the phage DNA, and *att*B, the bacterial attachment site normally present on the *E. coli* genome (see Fig. 4-6). It turns out that only *att*P must be present on a negatively supercoiled DNA, and that this requirement results from the particular way *att*P is wrapped around the protein core within the nucleoprotein complex at *att*P that performs DNA strand exchanges between *att*P and *att*B.[34] Interaction between *att*B and this nucleoprotein complex is not significantly dependent on DNA supercoiling; if *att*B is present on a separate DNA ring, the ring can be either relaxed or supercoiled in one integrase-mediated recombination.

Nature also utilizes the topological state of DNA to distinguish site-specific recombination between sites on the same DNA or on different DNA molecules, or to select properly oriented sites so as to control the outcome of the reaction.[35,36] As an example, consider the site-specific recombination promoted by an enzyme termed γδ resolvase. This enzyme normally promotes recombination between a pair of "direct repeats" on the same double-stranded DNA ring, and intermolecular recombination, or intramolecular recombination between two "inverted repeats" rarely occurs in this reaction (see the legend to Fig. 9-2 and the shaded panel on p. 178 for definitions of direct and inverted repeats).

How does a recombinase distinguish the orientations of its recognition sites and determine whether the sites are present on the same or different DNA molecules? It turns out that the substrate specificity in the γδ resolvase–catalyzed reaction is very closely tied to the topological state of the DNA. In Figure 9-2, a double-stranded DNA ring bearing two γδ resolvase recognition sites is represented by a closed line, and the

33. Reviewed in Felsenfeld, G., et al. 2000. *Biophys. Chem.* **86:** 231–237. See also Bancaud, A., et al. 2007. *Mol. Cell* **27:** 135–147.

34. Reviewed in Nash, H.A. 1990. *Trends Biochem. Sci.* **15:** 222–227.

35. Craigie, R. and Mizuuchi, K. 1986. *Cell* **45:** 793–800.

36. Reviewed in Gellert, M. and Nash, H. 1987. *Nature* **325:** 401–404; Grindley, N.D., et al. 2006. *Annu. Rev. Biochem.* **75:** 567–605.

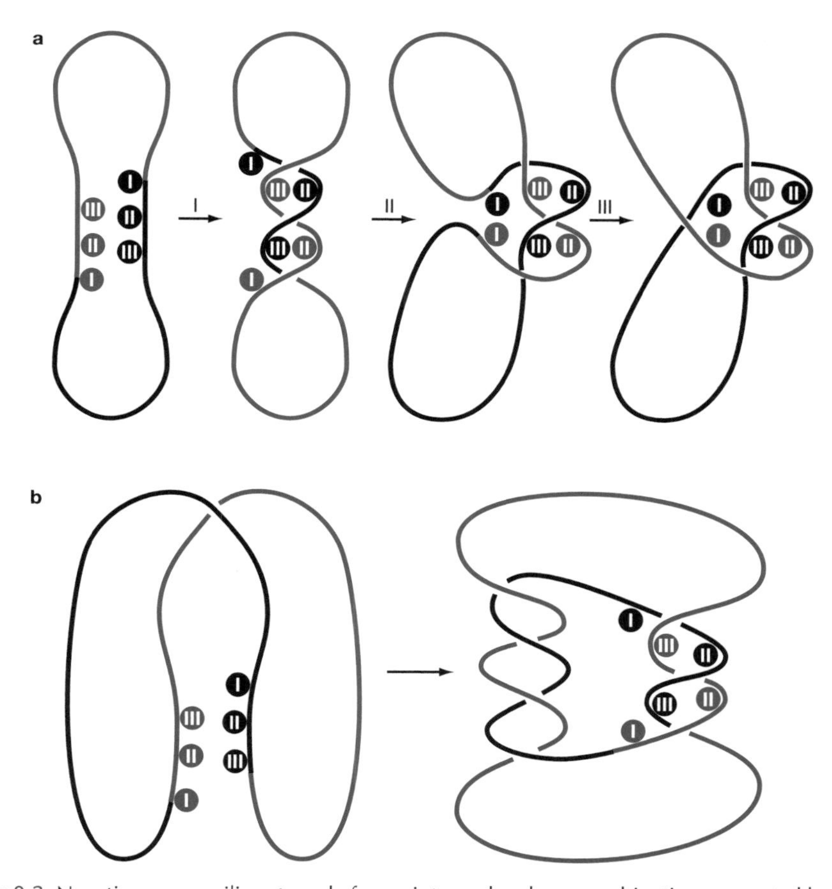

Figure 9-2. Negative supercoiling strongly favors intramolecular recombination promoted by γδ resolvase between two tandem sites rather than between two inverted sites. The line represents a double-stranded DNA ring, and the Roman numerals I, II, and III represent three short sequence blocks that constitute a recognition site of the resolvase. (*a*) The two sites are tandem or direct repeats of each other: Further along the DNA ring from one site, in the direction of I, II, and III, lies the second site, with the sequence blocks in the same order of I, II, and III. The right-handed intertwining of the γδ resolvase recognition sites in the recombinosome structure is responsible for the strong dependence of the γδ resolvase–promoted recombination on DNA negative supercoiling. See the text for details. (*b*) Here the two recognition sites are inverted rather than direct repeats of each other. Right-handed intertwining of the γδ resolvase recognition sites in a recombinosome structure is accompanied by left-handed wrapping elsewhere between the two DNA arcs; left-handed intertwining is disfavored in a negatively supercoiled DNA. (Adapted, with permission, from Grindley, N.D., et al. 2006. *Annu. Rev. Biochem.* **75:** 567–605, ©National Academy of Sciences U.S.A.)

In nucleic acid jargon, the term "direct repeats" refers to tandem copies of short stretches of identical or nearly identical nucleotide sequences along a DNA, and a pair of inverted repeats refers to two identical or nearly identical sequence elements that are oppositely oriented along a DNA. For example, in a DNA with the nucleotide sequence ······5′-GCAATCCG-3′······5′-GCAATCCG-3′······, the two tandem copies of the sequence 5′-GCAATCCG-3′ (and its complementary strand of the sequence 5′-CGGATTGC-3′, usually omitted to make the notation less cumbersome) constitute a direct repeat. The same sequence blocks in ······5′-GCAATCCG-3′······5′-CGGATTGC-3′······ constitute an inverted repeat. Here the two sequence blocks along the same DNA strand are complementary to each other, and, when viewed in their double-stranded forms along the duplex DNA, one duplex block is related to the other by a 180° rotation around an axis perpendicular to the DNA helical axis.

region marked I, II, and III represents three short sequence elements within each recognition site. The juxtaposition of a pair of recognition sites divides the DNA ring into two arcs, represented by lines of different colors in the figure.

Various experiments indicate that the resolvase first brings two copies of oppositely oriented II and III sequence blocks together, to align the two copies of sequence element I within which DNA strand exchange occurs (Fig. 9-2a, leftmost and the two middle drawings). As depicted in the two middle drawings in Figure 9-2a, in the resolvase–DNA nucleoprotein complex (a "recombinosome"), the two copies of the (II + III) sequence blocks are first intertwined in a right-handed way about each other. Because of this intertwining, strand exchange within the two copies of I would produce two singly linked DNA rings, one derived from the blue-colored arc and the other from the black arc (Fig. 9-2a, right), which are then unlinked by a type II DNA topoisomerase.

A consequence of the right-handed intertwining between the pair of II and III sequence blocks is that γδ recombinosome formation is strongly facilitated by negative supercoiling of the DNA ring. On the other hand, if the two resolvase sites are located on different molecules, then right-handed braiding between them would require formation of compensatory left-handed braids elsewhere in the pair of rings, which would be disfavored in a negatively supercoiled DNA. Furthermore, if the two sites are inverted rather than direct repeats on a DNA, then synapse between the (II + III) blocks would divide the DNA ring into two loops, one represented by the blue line, the other by the black line (Fig. 9-2b). Again, right-handed intertwining between these sequence blocks would require left-handed intertwining between the two loops, which would be disfavored in a negatively supercoiled DNA. It is such topological filtering that enables γδ resolvase to preferentially promote intramolecular recombination between two tandem site over intramolecular recombination between two inverted repeats, or intermolecular recombination.

DNA Supercoiling and Homologous Recombination

DNA supercoiling is utilized not only in promoting and controlling site-specific recombination, it may also have a significant role in general or homologous recombination—that is, recombination between DNA molecules with homologous sequences. Since the 1970s, supercoiling in bacterial DNA has been known to promote the formation of a recombination intermediate termed a "D-loop," in which a DNA strand "invades" a recipient double-stranded DNA to pair with the strand with a sequence complementary to it and displace the original partner of the invaded strand. For example, the *E. coli* RecA protein, a key player in homologous recombination, has been shown to promote D-loop formation between a DNA strand and a negatively supercoiled DNA.[37]

In eukaryotes, a major homologous recombination pathway involves a group of proteins termed the Rad52 group, which includes the Rad51, 52, and 54 proteins. In the presence of ATP, the human RAD54 protein hRAD54 can track along a DNA to generate a supercoiled loop.[38] A similar reaction mediated by the yeast Rad52 family members Rdh54 and Rad51 was also observed.[39] These findings suggest a translocation mechanism in which a positively supercoiled and a negatively supercoiled loop are simultaneously generated (Fig. 9-3), resembling the case in transcriptional supercoiling.[38,39] The formation of a negatively supercoiled loop may in turn facilitate pairing between the strand complementary to an invading strand (Fig. 9-3).[38–40]

A DNA CHAIN MAILLE

In contrast to strong evidence that Nature has turned the supercoiling of DNA into a powerful tool in its collection, few other consequences of DNA entanglement have provided equally convincing examples. The catenation and knotting of DNA rings and loops remain largely problems that Nature must resolve, rather than features it might utilize. Nevertheless, there are plausible exceptions to this generalization, and the mitochondrial DNA of a group of flagellated unicellular protozoa, called the kinetoplastids, might be one such example.

The kinetoplastids include both free-living and parasitic species that share the presence of a disk-shaped compact structure, termed the kinetoplast, within the single mitochondrion located at the base of the flagellum of each organism.[41] Studies of these

37. See, for example, Shibata, T., et al. 1979. *Proc. Natl. Acad. Sci.* **76:** 1638–1642; for earlier studies see the review by Radding, C.M. 1978. *Annu. Rev. Biochem.* **47:** 847–880.

38. Ristic, D., et al. 2001. *Proc. Natl. Acad. Sci.* **98:** 8454–8460; Sigurdsson, S., et al. 2002. *J. Biol. Chem.* **277:** 42790–42794.

39. Chi, P., et al. 2006. *J. Biol. Chem.* **281:** 26268–26279.

40. Kwon, Y., et al. 2007. *DNA Repair* **6:** 1496–1506.

41. For a review, see Lukes, J., et al. 2005. *Curr. Genet.* **48:** 277–299.

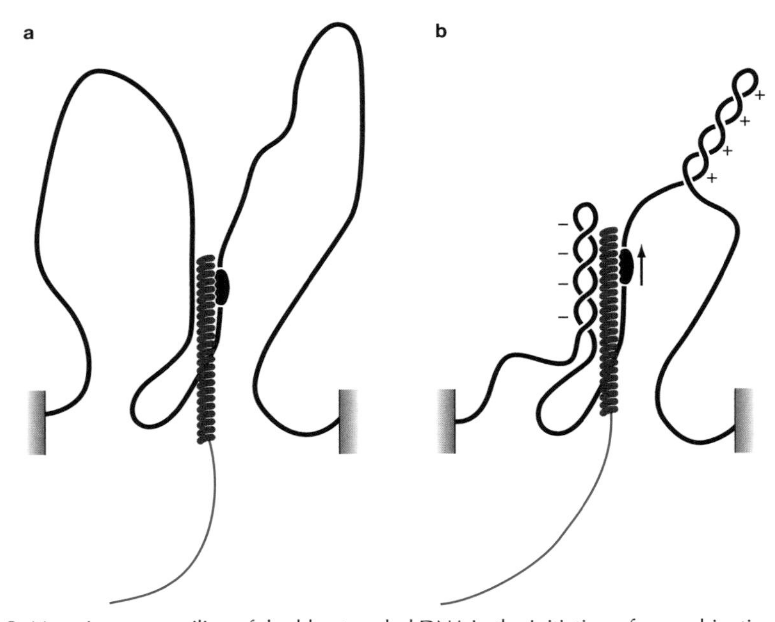

Figure 9-3. Negative supercoiling of double-stranded DNA in the initiation of recombination. Double-stranded DNA is represented by the black line and single-stranded DNA is represented by a blue line. (*a*) A loop of double-stranded DNA (between the two shaded areas) is invaded by a single-stranded DNA strand with a 3′ end. The nucleotide sequence of the single-stranded DNA is complementary to one strand of the duplex DNA. The 3′-terminal region of the invading strand is coated by a cluster of a RAD51 protein complex to form a helical filament. The kidney shape represents a DNA-bound RAD54 protein complex, which effects the translocation of the DNA double helix relative to the RAD54 protein complex. (*b*) Because the DNA-bound RAD54 protein complex also interacts with the RAD51 nucleoprotein filament, its rotational motion is restrained and the relative translocation between the complex and the double-stranded DNA leads to positive supercoiling of the DNA ahead of it and negative supercoiling of the DNA behind it (the arrow next to the RAD54 complex indicates the direction of its translocation relative to the DNA). The negative supercoiling of the DNA in turn facilitates the invasion of the double-stranded DNA by the RAD51-coated DNA strand, which forms a D-loop in the duplex DNA to initiate homologous recombination. (Adapted, with permission, from Ristic, D., et al. 2002. *Proc. Natl. Acad. Sci.* **98:** 8454–8460, ©National Academy of Sciences U.S.A.)

organisms, especially of the trypanosomatid subgroup, have been stimulated by findings that they cause a number of diseases in tropical and subtropical regions, including African sleeping sickness, Chagas' disease, and leishmaniasis. The disk-shaped kinetoplast is made of DNA rings that constitute the mitochondrial genome of the protozoan. The organization of these rings is one of the most bizarre masterpieces of Nature.

First, there are two kinds of rings, one that contains mainly sequences encoding several mitochondrial rRNAs and proteins, and the other with sequences encoding special RNAs termed "guide RNAs" or gRNAs, some of which are also encoded by the first kind of rings. The transcripts derived from those encoding rRNAs and proteins on the first type of rings are extensively encrypted, owing to the presence of extra Us at many places and the absence of needed Us at many other places. These encrypted transcripts therefore must be edited before they become functional, and the editing process, using the gRNA sequences encoded by both classes of DNA rings as templates, removes the superfluous Us or adds the missing Us. Thus, the two types of DNA rings form a functional set only in combination: One provides the encrypted information, and the other provides most of the decryption keys.

Second, in the kinetoplasts of the trypanosomes, the two types of rings form a complex catenated network resembling a medieval coat of chain maille made of interlinked metal rings.[42] The kinetoplast (kDNA) of *Trypanosoma brucei*, for example, consists of a network of a few dozen 23-kb rings of the first type and several thousand copies of 1-kb rings of the second type (Fig. 9-4). These two types of large and small rings have been dubbed the "maxicircles" and "minicircles," respectively. Whereas the maxicircles of a particular trypanosome have a unique nucleotide sequence, the minicircles contain several different sequence classes. An examination of 25 minicircles from *T. brucei* having differing sequences indicates that they encode 86 putative gRNAs, with an average of 2–5 gRNA per ring; in several cases, minicircles that are otherwise distinct sequences have been found to encode the same gRNAs.[43] It appears that the minicircles are interlocked to form a sheet, with each minicircle linked to an average of three other rings.[44] Inside a kinetoplastid, the rows of minicircles within a catenated sheet presumably can be pleated to form the compact disk-shaped kinetoplast, with a diameter of about 1 μm and a thickness of a few tenths of a μm (Fig. 9-4a).[44] The maxicircles also form a catenated network that is itself topologically linked with the minicircles network, at least in the case of *Trypanosoma equiperdum*, the pathogen that causes a sexually transmitted equine disease called dourine.[45]

Studies of *T. brucei* and *Crithidia fasciculata* kDNA have revealed a number of fascinating features of minicircle replication.[46,47] Individual minicircles with intact DNA strands are first unlinked from the network by a type II DNA topoisomerase, presumably the mitochondrial enzyme Top2mt,[48] and are then replicated in a zone

42. Riou, G. and Delain, E. 1969. *Proc. Natl. Acad. Sci.* **64:** 618–625.

43. Hong, M. and Simpson, L. 2003. *Protist* **154:** 265–279.

44. Chen, J., et al. 1995. *Cell* **80:** 61–69; Chen, J., et al. 1995. *EMBO J.* **14:** 6339–6347.

45. Shapiro, T.A. 1993. *Proc. Natl. Acad. Sci.* **90:** 7809–7813. There is evidence that strains termed *T. equiperdum* and *T. evansi,* which cause diseases in livestock, are actually mutated *T. brucei* with a nonfunctional kinetoplast DNA (see Lai, D.H., et al. 2008. *Proc. Natl. Acad. Sci.* **105:** 1999–2004).

46. Liu, B., et al. 2005. *Trends Parasitol.* **21:** 363–369; Lukes, J., et al. 2002. *Eukaryot. Cell* **1:** 495–502.

47. Shapiro, T.A. and Englund, P.T. 1995. *Annu. Rev. Microbiol.* **49:** 117–143; Shlomai, J. 2004. *Curr. Mol. Med.* **4:** 623–647.

48. See, for example, Melendy, T., et al. 1988. *Cell* **55:** 1083–1088; Shapiro, T. A., et al. 1989. *J. Biol. Chem.* **264:** 4173–4178; Kulikowicz, T. and Shapiro, T.A. 2005. *J. Biol. Chem.* **281:** 3048–3056.

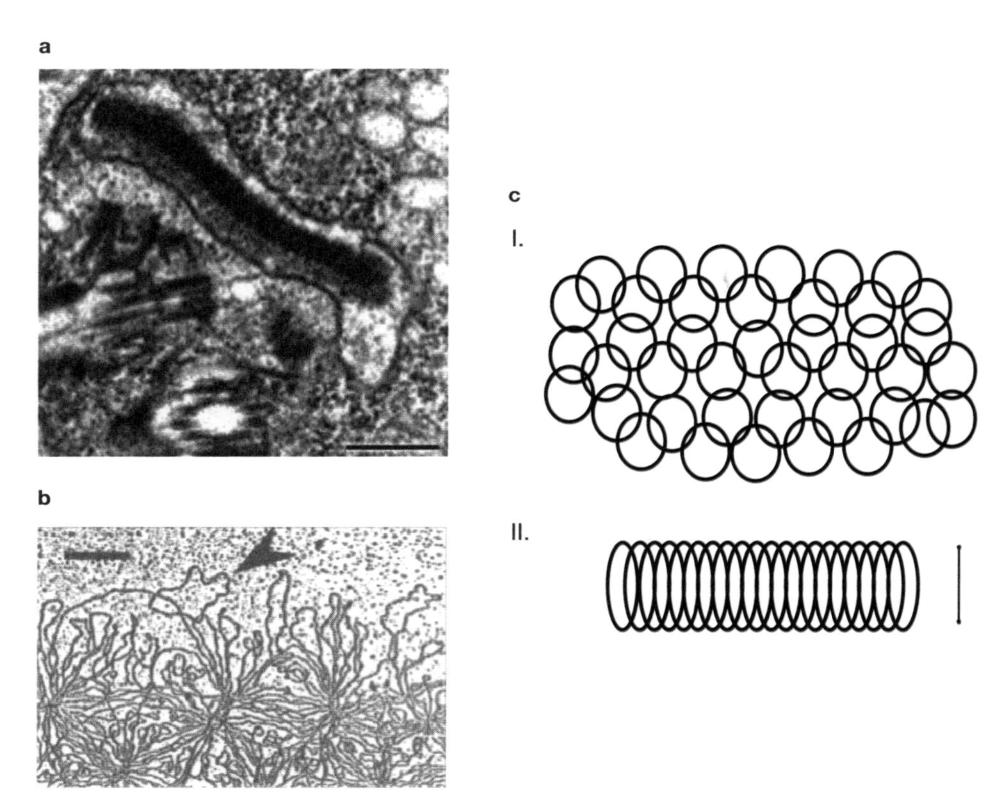

Figure 9-4. The kinetoplast of *Trypanosoma brucei*. (a) Electron micrograph of a cross section of a region of a *T. brucei* protozoan containing the kinetoplast. In this cross section, the disk-shaped kDNA is seen as a dark rod inside the single mitochondrion, and the bar in the *lower right* corner represents a length of 0.25 μm. The kDNA disk shown here is in a late stage of DNA replication, and hence its diameter (the length of the rod) has expanded relative to that of a nonreplicating kDNA. (From the work of Ogbadoyi, E.O., et al. 2003. *Mol. Biol. Cell.* **14:** 1769–1779) (*a:* Reprinted, with permission of Elsevier, from Liu, B., et al. 2005. *Trends Parasitol.* **21:** 363–369.) (*b*) The image shows an electron micrograph of the periphery of purified *Trypanosoma avium* kDNA network. The bar at the *upper left* corner marks a length of 0.5 μm. The loops in the micrograph represent interlinked minicircles; one that is clearly discernible is marked with an arrowhead. (*c*) Schematic drawings illustrate a section of isolated kDNA network with interlinked minicircles (I), and a section of the condensed network disk (II) inside the trypanosome mitochondrion. (Part *b* is reprinted and part *c* is adapted, with permission of the American Society for Microbiology, from Lukes, J., et al. 2002. *Eukaryot. Cell* **1:** 495–502.)

between the flagellar basal body and the compact structure containing the kDNA network. The replicated rings, in a form that has at least one single-stranded nick or gap, are then returned to two antipodal sites flanking the disk-shaped body of DNA rings and reattached to the network periphery.[46] The DNA topoisomerase responsible for reattachment could be either a type II enzyme, such as Top2mt, or

Top1mt,[49] a type IA enzyme found in kinetoplastid mitochondrion. On the basis of the known characteristics of other type IA DNA topoisomerases (discussed in Chapter 3), Top1mt is expected to be capable of attaching nicked or gapped rings, but not rings with intact strands, to a catenated DNA network.

Replicated rings are converted to the covalently closed form only after all minicircles in a catenated network have been replicated.[44] Thus, the kinetoplast replication system appears to utilize a difference between unreplicated DNA and replicated rings to distinguish them. This difference could be topological—that is, the topological difference between a ring with a single-stranded nick or gap and a ring that has intact strands (see our earlier discussions of DNA supercoiling). But it is equally plausible that this determining difference could be the structural features of a nick or gap that mark the replicated rings, or perhaps the presence of a one-use-only macromolecular tag on an unreplicated ring—a single-use "licensing factor."

Why is a kDNA organized in a catenated network containing thousands of DNA rings? One suggestion is that this network provides a system, primitive but nevertheless serviceable, for transmitting a full complement of kinetoplast-encoded sequences during cell division.[50] In general, a full complement of kinetoplast-encoded sequences must contain at least one maxicircle and a minimal number of minicircles encoding a complete set of all gRNA sequences not encoded by the maxicircle. Clearly, a certain device is necessary to organize and orchestrate the division and partition of such a collection, to ensure that none of the DNA rings is lost from one generation to the next. These kinetoplasts may have found one way of reducing the chance of any genetic loss by topologically linking the multiple genetic elements into a network of rings. Nevertheless, it is also possible that the generalization that "Nature does nothing useless" is not without exceptions. Even among the kinetoplastids—including *Bodo caudatus*, *Bodo saltans*, *Trypanoplasma borreli*, and *Cryptobia helices*—we see that some seem to manage quite well without a catenated kDNA.[51]

49. Scocca, J. R. and Shapiro, T. A. 2008. *Mol. Microbiol.* **67:** 820–829.

50. Borst, P. 1991. *Trends Genet.* **7:** 139–141.

51. Hajduk, S. L., et al. 1986. *Mol. Cell. Biol.* **6:** 4372–4378; Yasuhira, S. and Simpson, L. 1996. *RNA* **2:** 1153–1160; Lukescaron, J., et al. 1998. *EMBO J.* **17:** 838–846; Blom, D., et al. *RNA* **6:** 121–135.

CHAPTER 10

From Nature's Battlefields to the Clinical Wards

"The doctrine that all nature is at war is most true."
Charles C. Darwin, in *Natural Selection*

THERE ARE CONSTANT BATTLES IN THE BIOSPHERE. Those being fought with teeth and muscle in the African plains, in the Amazon rain forests, or in the deep blue seas are vividly presented for the naked eyes to see. Yet there are equally ferocious battles, everywhere and at all times, that are invisible and silent. The main weapons used in those engagements are often of a chemical and molecular nature, and the strategies invariably ingenious and intricate. The eternal war between a parasitic F plasmid and its host bacterium *Escherichia coli* serves as a good example.

A successful parasite is one that does not kill its host, but remains tenacious in that it never leaves its host. The F plasmid, also called the F factor, is a ring-shaped DNA of some 60,000 bp. There is a single copy of the plasmid in each *E. coli* host cell, and it divides once per division cycle of its host. To ensure its success as a parasite, the F plasmid encodes several proteins that are dedicated to the equal partition of the two F DNA progenies into a pair of newly duplicated *E. coli* cells, so that each newborn *E. coli* from an F-bearing parent is destined to inherit one F plasmid.

The plasmid partition system is not entirely fail-safe, however. At a low frequency, one of a pair of *E. coli* progeny may become free of the F plasmid whereas its twin inherits both copies of the duplicated F plasmid. In such a case, the parasite has another brutal and nasty trick in its repertoire to ensure that the freedom enjoyed by the plasmid-free *E. coli* progeny is short: The cell without an F plasmid is soon killed to prevent it from producing parasite-free descendants.

How does an F factor accomplish this feat in its absence? We now know that the F factor utilizes a toxin–antitoxin strategy.[1] The plasmid encodes two proteins, CcdA (also called LetA) and CcdB (also called LetD). CcdB is a toxin and CcdA an antitoxin. Together, the pair forms a complex that is harmless to the host cell, but in the absence of CcdA, CcdB is lethal to the cell. Furthermore, whereas CcdB is fairly stable, CcdA is labile and must be constantly replenished in *E. coli* cells bearing the F plasmid. Thus, if a cell derived from an F-bearing parent loses the plasmid, it can no longer replenish its decaying CcdA, and the longer lasting CcdB toxin inherited from the parental cytoplasm seals its fate.

How does CcdB kill *E. coli*? CcdB destroys *E. coli* by hijacking DNA gyrase (topoisomerase II) and turning it into a DNA-damaging agent. CcdB is a protein composed of two identical peptides, each of 101 amino acid residues. If an adequate amount of the antitoxin CcdA is present, it forms a stable complex with CcdB. But in the absence of CcdA, CcdB forms a complex with a DNA-bound gyrase. In the CcdB–gyrase complex, the DNA gate in the gyrase–DNA complex is trapped in its unlocked state.[1] Thus the CcdB toxin prevents the rejoining of the gyrase-linked DNA in the CcdB–gyrase–DNA ternary complex. Results of X-ray diffraction of the toxin and mutagenesis studies of *E. coli* mutants resistant to it suggest a remarkable model of the CcdB–gyrase–DNA ternary complex, in which CcdB docks into the large central hole in gyrase with the DNA gate open (Fig. 10-1).[2] In this ternary complex, the enzyme can no

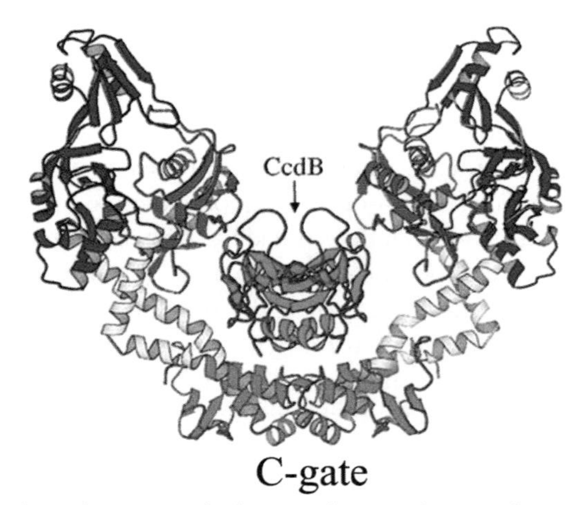

Figure 10-1. Docking of a CcdB toxin inside the central cavity of a GyrA dimer in its "open" conformation. The open conformation is believed to resemble the gyrase conformation after it has cleaved a bound DNA double helix and pulls apart the two enzyme-attached DNA ends. The CcdB toxin is shown in blue. (Reprinted, with permission of Elsevier, from Loris, R., et al. 1999. *J. Mol. Biol.* **285:** 1667–1677.)

1. Reviewed in Couturier, M., et al. 1998. *Trends Microbiol.* **6:** 269–275; Kamphuis, M.B., et al. 2007. *Protein Pept. Lett.* **14:** 113–124; see also Miki, T., et al. 1992. *J. Mol. Biol.* **225:** 39–52; Bernard, P. and Couturier, M. 1992. *J. Mol. Biol.* **226:** 735–745; Bernard, P., et al. 1993. *J. Mol. Biol.* **234:** 534–541.

2. Loris, R., et al 1999. *J. Mol. Biol.* **285:** 1667–1677; Dao-Thi, M. H., et al. 2005. *J. Mol. Biol.* **348:** 1091–1102.

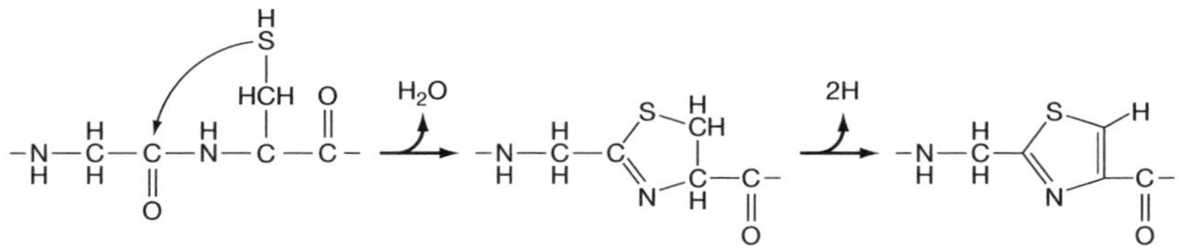

Figure 10-2. Formation of a thiazole ring from an adjacent pair of glycyl and cysteinyl residues in the 43-amino-acid-long microcin B17 precursor. In a similar reaction, a seryl hydroxyl rather than a cysteinyl sulfhydryl reacts with an adjacent glycyl residue to form an oxazone ring, in which an oxygen replaces the sulfur atom in the thiazole. All of these reaction steps are catalyzed by enzymes encoded by plasmids expressing the toxin. (Redrawn, with permission, from Yorgey, P., et al. 1994. *Proc. Natl. Acad. Sci.* **91:** 4519–4523, ©National Academy of Sciences U.S.A.)

longer rejoin the broken DNA, and the presence of the DNA break eventually leads to cell death (as we shall see below).

In a strategy similar to that used by the F plasmid, several other *E. coli* plasmids also encode a toxin called microcin B17 (MccB17), as well as its antitoxin. Like CcdB, MccB17 also targets gyrase and traps the covalent gyrase–DNA complex.[3] In contrast to CcdB, however, the MccB17 toxin undergoes a series of transformations. The 69-amino-acid-long MccB17 precursor is first trimmed by the removal of its amino-terminal 26 residues, and the shortened peptide is then extensively processed by several plasmid-encoded enzymes to yield a product that barely resembles a peptide. The final product contains four thiazole and four oxazone rings, formed from adjacent glycyl and cysteinyl and glycyl and seryl residues, respectively (Fig. 10-2).[4]

DNA TOPOISOMERASE-TARGETING THERAPEUTICS FROM NATURAL SOURCES

CcdB and MccB17 are only two of a plethora of DNA topoisomerase-targeting weapons in Nature's arsenal. Shortly after the discovery of DNA gyrase in 1976, the enzyme was identified as the target of two coumarin antibiotics, coumermycin A_1 and novobiocin, that are synthesized in certain soil bacteria of the genus *Streptomyces*.[5] Other than these coumarins, many chemically diverse natural products of bacterial and plant origins, including cinodines, flavones, and terpenoids, have been found to target bacterial type II DNA topoisomerases.

The identification in the 1980s of mammalian DNA topoisomerases I and II as the targets of a large number of anticancer drugs greatly expanded interest in these en-

3. Vizan, J.L., et al. 1991. *EMBO J.* **10:** 467–476.
4. Yorgey, P., et al. 1994. *Proc. Natl. Acad. Sci.* **91:** 4519–4523; Li, Y.M., et al. 1996. *Science* **274:** 1188–1193.
5. Gellert, M., et al. 1976. *Proc. Natl. Acad. Sci.* **73:** 4474–4478.

zymes as drug targets.[6] Many of the anticancer drugs were isolated or derived from natural sources, often without any prior knowledge of their cellular targets. Among the DNA topoisomerase II- or Top2-targeting therapeutics, the anthracycline drugs daunorubicin, doxorubicin, and epirubicin are derived from the fungus *Streptococcus peucetius*. The aristolochic acid derivative amonafide, and the epipodophyllotoxins etoposide and teniposide (also called VP-16 and VM-26, respectively), are all plant metabolites. Similarly, the DNA topoisomerase I- or Top1-targeting drugs irinotecan and topotecan are derivatives of camptothecin, a plant alkaloid. Whereas the biological function of the plasmid-encoded toxins CcdB and MccB17 is clear, the plausible biological roles of the parent compounds of many other topoisomerase-active substances are yet to be established. It is most likely that many if not all of them are also involved in Nature's warfare.

DNA TOPOISOMERASE-TARGETING THERAPEUTICS FROM CHEMICAL SYNTHESIS

A very large number of compounds have been synthesized in chemical laboratories and tested for their potential in fighting diseases. At the time of the discovery of gyrase in 1976, the chemically synthesized antibiotic nalidixic acid was in wide use in the treatment of urinary infections. This drug and its analog, oxolinic acid, which is used in veterinary medicine, were then known to inhibit DNA replication in *E. coli*. The question was therefore raised of whether gyrase, which had been identified as the target of the replication inhibitors novobiocin and coumermycin, could also be the target of nalidixic and oxolinic acid. It turns out that the enzyme is indeed the target of these antibiotics as well. But whereas coumarin antibiotics act on the B-subunit of gyrase, nalidixic and oxolinic acid mainly target the A-subunit of the enzyme.[7]

Nalidixic and oxolinic acid, developed in the early 1960s, are first-generation members of what has been termed the quinolone class of antibiotics. The discovery of gyrase as their target has greatly accelerated the development of more potent compounds of this class, and a long list of new entrants (see Table 10-1) has been developed over the years.[8] Many of the newer quinolones have a fluorine atom at the sixth carbon in the basic quinolone structure (Fig. 10-3), and these compounds are often referred to as the fluoroquinolones. After the discovery of bacterial DNA topoisomerase IV (Top4) in 1990, it became known that the quinolone antibiotics may act on either

6. Reviewed in Liu, L.F. 1989. *Annu. Rev. Biochem.* **58:** 351–375.

7. Sugino, A., et al. 1977. *Proc. Natl. Acad. Sci.* **74:** 4767–4771; Gellert, M., et al. 1977. *Proc. Natl. Acad. Sci.* **74:** 4772–4776. Subsequently, both GyrA and GyrB subunits were shown to contribute to the binding of these drugs; hence the use of the adverb "mainly" in the statement that "nalidixic and oxolinic acid mainly target the A subunit of the same enzyme."

8. See, for example, the reviews by Maxwell, A. 1997. *Trends Microbiol.* **5:** 102–109; Hooper, D.C. 1998. *Clin. Infect Dis.* **27:** S54–S63; Mitscher, L.A. 2005. *Chem. Rev.* **105:** 559–592.

Table 10-1. Examples of drugs that target the DNA topoisomerases

Target	Drug	Binding site
Mammalian DNA topoisomerase I	Camptothecin derivatives: topotecan, irinotecin (CPT-11)	Site of DNA cleavage
Mammalian DNA topoisomerase II	Daunorubicin (daunomycin), doxorubicin (adriamycin), mitoxantrone, etoposide, teniposide	Site of DNA cleavage
Bacterial type IIA DNA topoisomerases	Quinolone class: nalidixic acid, ciprofloxacin, norfloxacin, enoxacin, levofloxacin, norfloxacin, sparfloxacin, trovafloxacin	Site of DNA cleavage
Mammalian type IIA DNA topoisomerases	Bisdioxopiperazine class: dexrazoxane (ICRF-187), ICRF-154, ICRF-159, ICRF-193, MST-16	ATPase domain

of the two bacterial type IIA DNA topoisomerases; the primary target of a particular compound may be gyrase or Top4, depending on the particular bacterium involved.[8]

Efforts in synthetic chemistry have also led to an ever-lengthening list of DNA topoisomerase-targeting compounds with antitumor activities. Among these, the anthracenedione derivative mitoxantrone has proven clinically useful, and another derivative, amsacrine, is sometimes used in combination with other anticancer agents.[9] Both of these compounds act on the type II DNA topoisomerases.[6,10] A class of Top2-targeting compounds known as bisdioxopiperazines (also called bisdiketopiperazines or razopiperazines), initially studied in the late 1960s for their antitumor activity and

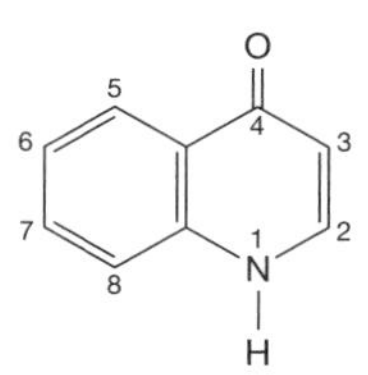

Figure 10-3. The basic chemical structure of the quinolones. The carbon atoms in the double ring structure and the hydrogen atoms attached to them are omitted for brevity. In some of the quinolones, position 8 of the ring is also occupied by a nitrogen instead of a carbon. Nalidixic acid, for example, is a quinolone with a nitrogen at both positions 1 and 8, a methyl group (—CH_3) at the N1 as well as at the C7 position, and a carboxyl group (—COOH) at the C3 position. A number of the quinolone antibiotics, including norfloxacin, ciprofloxacin, and sparfloxacin, have a fluorine atom at position 6, and these agents are collectively called fluoroquinolones.

9. Cassileth, P.A. and Gale, R.P. 1986. *Leuk. Res.* **10:** 1257–1265; Koeller, J. and Eble, M. 1988. *Clin. Pharm.* **7:** 574–581.

10. Kreuzer, K.N. 1998. *Biochim. Biophys. Acta* **1400:** 339–347.

later found to target Top2, are also of substantial interest.[11] Among them, the compound ICRF-187 (dexrazoxane) has been used to counter cardiotoxicity caused by the Top2 drug doxorubicin.[12]

MODES OF ACTION OF DRUGS TARGETING THE DNA TOPOISOMERASES

Some of the anticancer drugs that have proven clinically useful in targeting DNA topoisomerases are listed in Table 10-1. Whereas some of these compounds are closely related—such as topotecan and irinotecan, both derivatives of camptotecan—others show a broad chemical diversity despite their sharing of the same molecular targets. The lack of common structural features among drugs targeting the same enzyme is partly due to the presence of multiple structural domains in the target enzyme; thus, a single enzyme has several chemically distinct sites for drug binding. Furthermore, during the catalytic cycle of a DNA topoisomerase, the enzyme–DNA complex assumes a number of distinct conformations. These dynamic changes provide another dimension for drug action, some of which are discussed in the following sections.

Trapping of Covalent Enzyme–DNA Complexes

The great majority of the many DNA topoisomerase-targeting drugs now in clinical use act by trapping covalent enzyme–DNA complexes (see Table 10-1). Examples are the Top1-targeting camptothecin derivatives topotecin and irinotecan, a large number of Top2-targeting drugs including the quinolone antibiotics, and the anticancer therapeutics doxorubicin, etoposide, and mitoxantrone.

Whereas all of these compounds stabilize covalent topoisomerase–DNA complexes, they may do so in different ways. We have seen that CcdB, a bacterial plasmid-encoded toxin, traps the gyrase–DNA covalent complex not by directly binding to a region close to the DNA cleavage sites of the enzyme, but rather by landing inside the central cavity above the gyrase C-gate region where the T-segment normally makes its exit after sailing through the DNA gate. Because of the conformational coupling among various parts of the enzyme–DNA complex, binding of CcdB to the C-gate region apparently interferes with the closing of the DNA gate.

By contrast, many other agents that trap various topoisomerase–DNA covalent complexes bind directly to regions proximal to the sites of DNA cleavage. A well-studied case in point is the action of camptothecin and its derivatives. Here X-ray crystallography reveals that in a human Top1–DNA–topotecan ternary complex, the planar polycyclic ring

11. Reviewed in Andoh, T. and Ishida, R. 1998. *Biochim. Biophys. Acta* **1400:** 155–171; see also Tanabe, K., et al. 1991. *Cancer Res.* **51:** 4903–4908.

12. Cvetkovic, R.S. and Scott, L.J. 2005. *Drugs* **65:** 1005–1024.

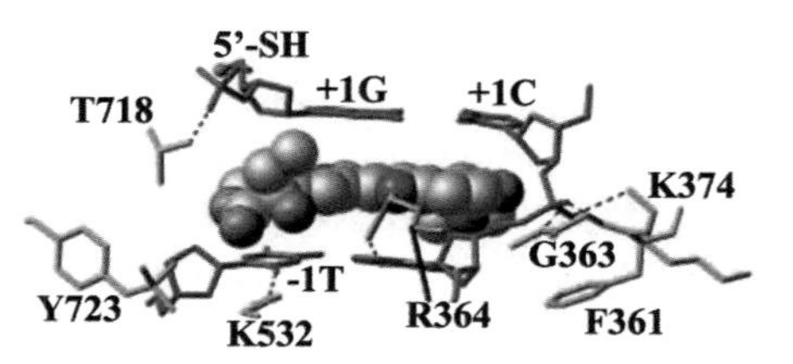

Figure 10-4. Insertion of the Top1-targeting anticancer drug topotecan in the Top1–DNA covalent complex. Only the –1 and +1 DNA base pairs and several of the amino acids (in one-letter amino acid codes) that are involved in stabilizing the Top1–DNA–topotecan ternary complex are shown in the illustration. The bulk of the DNA and protein are omitted for clarity. The red, blue, and gray spheres represent, respectively, the oxygen, nitrogen, and carbon atoms that constitute topotecan. The DNA bases are represented by blue sticks; the –1 T is covalently linked to Tyr-723 (Y723) of the enzyme, and a sulfhydryl group is present on the +1 G (pink sphere), instead of the normal hydroxyl group in the enzyme–DNA covalent complex, thus preventing the rejoining of the ends of the broken DNA strand. (Reprinted, with permission, from Staker, B.L., et al. 2002. *Proc. Natl. Acad. Sci.* **99:** 15387–15392, ©National Academy of Sciences U.S.A.)

system of the drug "intercalates," or inserts itself, between the DNA base pairs flanking the point of DNA strand disruption, as shown in Figure 10-4 (see Chapter 4 for enumeration of the DNA base pairs at various topoisomerase binding sites in DNA—it is the scissile-strand nucleotide at base pair –1 that becomes covalently linked to the enzyme during transesterification).[13] The inserted drug moiety thus displaces the downstream or +1 base pair in the normal covalent enzyme-DNA complex, increasing its separation from the -1 base pair by ~3.6 Å and reducing the twist angle between the +1 and –1 base pairs by ~26°. Consequently, the separation between the 5′-OH and the protein-attached 3′-phosphorus is changed from 3.5Å in the normal binary enzyme–DNA covalent complex to 11.5 Å in the drug–enzyme–DNA ternary complex (note that a 5′-SH rather than a 5′-OH is present in the ternary complex used for crystallization to prevent the rejoining of the DNA backbone bond; see Chapter 4). The large increase in distance between the DNA end groups normally posited for reformation of the DNA backbone bond readily explains how topotecan prevents the rejoining of the broken DNA strand.

The intercalation of the polycyclic ring system between the –1 and +1 base pairs is the most striking feature of the topotecan–Top1–DNA ternary complex. This feature is also seen in the ternary complexes of camptothecin itself and two members of the indocarbazole and indenoisoquinoline classes of DNA topoisomerase I "poisons," a term often used for drugs that trap topoisomerase–DNA covalent complexes and thus convert their target enzymes to DNA damaging and cytotoxic agents, the latter compounds being chemically very distinct from the camptothecins.[14] There are,

13. Staker, B.L., et al. 2002. *Proc. Natl. Acad. Sci.* **99:** 15387–15392.

14. Staker, B.L., et al. 2005. *J. Med. Chem.* **48:** 2336–2345.

however, additional characteristic interactions in the drug–enzyme–DNA ternary complexes. These include not only drug-enzyme contacts (Fig. 10-4), but also specific protein–DNA interactions that occur only in the presence of the individual drugs. All of these interactions are also important in determining drug specificity. The camptothecins, for example, are potent poisons to type IB enzymes represented by yeast and human DNA topoisomerase I, but have little effect on the mechanistically closely related type IB enzymes of the poxvirus topoisomerase type.

Although X-ray structural data are unavailable for DNA topoisomerase II therapeutic agents that trap covalent enzyme–DNA complexes, biochemical and genetic data suggest that a number of these agents also act near the DNA cleavage sites of their target enzymes. Several experiments support the notion that the structurally distinct Top2-targeting drugs etoposide, amsacrine, genistein, and CP-115,953 (a quinolone) have overlapping binding pockets in the drug–enzyme–DNA complexes.[15] For example, the bacterial gyrase poison ciprofloxacin, although ineffective in enhancing DNA cleavage by eukaryotic Top2, is a competitive inhibitor of both etoposide and CP-115,953 in their trapping of covalent complexes between eukaryotic DNA topoisomerase II and DNA.[16] A direct demonstration that amsacrine is located in or near the DNA cleavage pockets of its target enzyme has been accomplished through the use of a photo-activatible azido derivative of the drug.[17] The substituted amsacrine, which like its parent compound enhances DNA cleavage by the type II DNA topoisomerases, including phage T4 DNA topoisomerase, is shown to photo-crosslink to DNA bases immediately adjacent to the pair of phosphodiester bonds that are cleaved by the enzyme.[18] Another study examined photochemical cleavage of DNA in the vicinity of a fluoroquinolone analog in a drug–*Drosophila* Top2-DNA ternary complex. This cleavage, promoted by a photon-initiated free radical reaction at the fluoroquinolone, is found to occur at locations close to those introduced by the topoisomerase itself.[19] In such photo-crosslinking and photo-cleavage reactions, it is known that only DNA immediately adjacent to the photochemical reaction centers is affected. Therefore, these data provide strong evidence that the amsacrine derivative and the fluoroquinolone analog are located within or near the normal DNA cleavage pockets of DNA topoisomerase II.

Mutagenesis data from studies of quinolone resistance of bacterial type II DNA topoisomerases also support the notion that the quinolones bind to sites within or

15. Osheroff, N., et al. 1994. *Cancer Chemother. Pharmacol.* **34:** S19–25.

16. Elsea, S.H., et al. 1997. *Biochemistry* **36:** 2919–2924.

17. Upon irradiation with ultraviolet light, an azido group ($N = N^+ = N^-$) gives off a nitrogen molecule (N_2) to form a highly reactive nitrene (a nitrogen atom with six electrons around it), which can react with a nearby amino acid side chain to form a stable covalent bond.

18. Freudenreich, C.H. and Kreuzer, K.N. 1994. *Proc. Natl. Acad. Sci.* **91:** 11007–11011.

19. Yu, H., et al. 2000. *Biochemistry* **39:** 10236–10246.

near the DNA breakage-and-rejoining sites of their target DNA topoisomerases.[20] However, as we saw for the camptothecin class of compounds, resistance to a compound resulting from a change in an enzyme amino acid side chain is not limited to changes in amino acid side chains that interact directly with the compound. Furthermore, a change in amino acid at a particular location may sometimes cause a distal structural alteration in the enzyme. Thus, the mutant sites that confer drug resistance are not always indicative of the locations of a drug in a ternary complex.

Interfering with ATP Utilization

Novobiocin and coumermycin are the earliest known examples of topoisomerase-targeting drugs that interfere with ATP utilization by type II DNA topoisomerases.[20,21] Several crystal structures have been determined for this class of type II DNA topoisomerase inhibitors,[22] including the coumarin antibiotics novobiocin, coumermycin A_1, and clorobiocin, and the cyclothialidine GR122222X. GR122222X is a derivative of a natural product of *Streptomyces filipinensis*, with a chemical structure distinct from that of the coumarins. Despite their chemical dissimilarity, all of these gyrase inhibitors are found to bind to similar sites in the GyrB part of gyrase. The small but significant degree of overlap among the regions they occupy and the binding pockets of the ATP analog ADPNP (Fig. 10-5) explains why these drugs behave like competitive inhibitors of the ATPase activity of gyrase.

Locking the Entrance Gate

The bisdioxopiperazines that target DNA topoisomerase II,[11] including ICRF-187 (dexrazoxane) and ICRF-193 (see Table 10-1), appear to act by locking the ATP-modulated entrance gate of the enzyme (Fig. 10-6a).[23] In the presence of ATP, these compounds can bind to the closed-clamp conformation of the enzyme and prevent reopening of the entrance gate.[23] X-ray crystallography of an ICRF–187–Top2 complex shows that the drug binds and bridges the dimer interface between the two ATPase protomers, with the dyad axis of the drug molecule coinciding with that of the dimeric protein (Fig. 10-6b).[24] The bisdioxopiperazine drugs therefore do not compete for the

20. See, for example, Yoshida, H., et al. 1990. *Antimicrob. Agents Chemother.* **34:** 1271–1272; Yoshida, H., et al. 1991. *Antimicrob. Agents Chemother.* **35:** 1647–1650; Heddle, J. and Maxwell, A. 2002. *Antimicrob. Agents Chemother.* **46:** 1805–1815.

21. Sugino, A., et al. 1978. *Proc. Natl. Acad. Sci.* **75:** 4838–4842; Mizuuchi, K., et al. 1978. *Proc. Natl. Acad. Sci.* **75:** 5960–5963. For a review of drugs that bind to the ATPase domains of type II DNA topoisomerases, see Maxwell, A. and Lawson, D. M. 2003. *Curr. Top. Med. Chem.* **3:** 283–303.

22. Lewis, R.J., et al. 1996. *EMBO J.* **15:** 1412–1420; Tsai, F.T., et al. 1997. *Proteins* **28:** 41–52.

23. Roca, J., et al. 1994. *Proc. Natl. Acad. Sci.* **91:** 1781–1785.

24. Classen, S., et al. 2003. *Proc. Natl. Acad. Sci.* **100:** 10629–10634; Erratum in *Proc. Natl. Acad. Sci.* **100:** 14510.

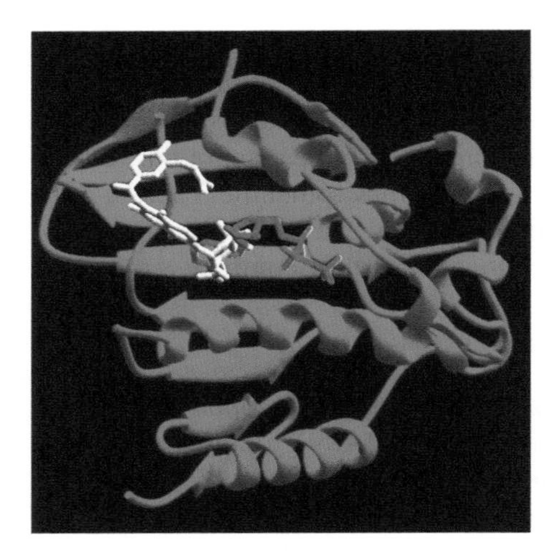

Figure 10-5. Overlap of the novobiocin and ATP binding sites in the ATP-binding domain of the *E. coli* GyrA subunit. The structure of the GyrA fragment is shown in a Ribbons representation, and the structure of a bound novobiocin (white) is superimposed on that of the bound nonhydrolyzable ATP analog ADPNP (blue) to show that the two binding sites overlap. (Reprinted, with permision of Macmillan Publishers Ltd., from Lewis, R.J., et al. 1996. *EMBO J.* **15:** 1412–1420.)

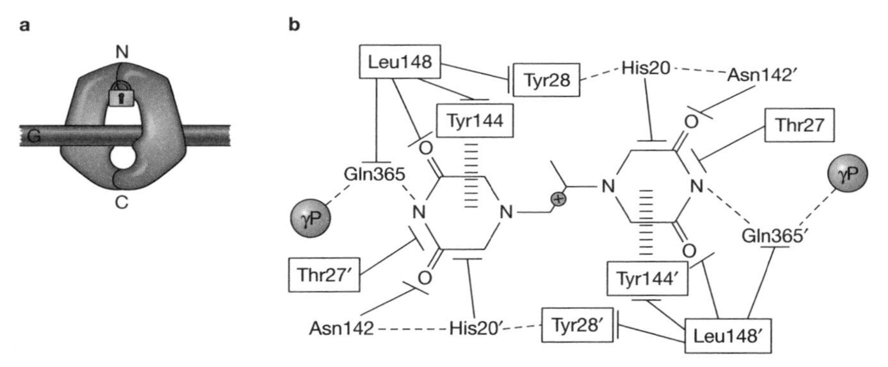

Figure 10-6. The action of a drug of the bisdioxopiperazine class. (*a*) Schematic illustration of the entrance and exit gates of a type IIA DNA topoisomerase, labeled by N and C, respectively, and the DNA G-segment, represented by a rod labeled G. The bisdioxopiperazine drug is represented by a padlock that locks the entrance gate upon its closure by ATP binding. (*b*) Schematic illustration of specific interactions between the bisdioxopiperazine drug ICRF-187 (dexrazoxane) and yeast DNA topoisomerase II in its closed-clamp conformation. Amino acid residues likely to be involved in binding dexrazoxane are indicated (those from one protomer of the dimeric enzyme are denoted with a prime [Leu148′, Tyr28′, etc.], and those without a prime are residues from the other protomer). Boxed residues are those used in making drug-resistant mutants. The pair of circles labeled γP mark the positions of the γ phosphates of the two ATP molecules bound to the enzyme. Dotted lines indicate hydrogen bonds; horizontal dashed lines indicate the stacking interactions between a piperazine ring and the phenyl ring in a tyrosyl residue; and solid lines with a flat end indicate interactions between uncharged atoms that are close to each other. The circle with a cross inside indicates the position of the molecular dyad of the enzyme, which is perpendicular to the plane of the paper; the drug is also symmetrical about this dyad except for the presence of a methyl group near the dyad. (Modified, with permission, from Classen, S., et al. 2003. *Proc. Natl. Acad. Sci.* **100:** 10629–10634, ©National Academy of Sciences U.S.A.)

ATP-binding pockets in a type II DNA topoisomerase, but rather stabilize the closed-clamp form of the enzyme either in the free or DNA-bound form, and prevent the enzyme from turning over.

Interfering with DNA Binding

The compound simocyclinone D8, produced by certain strains of *Streptomyces antibioticus,* is structurally related to the coumarin antibiotics and also targets DNA gyrase. However, whereas the coumarins inhibit gyrase by binding to the ATPase pockets of GyrB, and are therefore competitive inhibitors of the ATPase activity of the enzyme, simocyclinone D8 has no effect on the weak ATPase activity of gyrase in the absence of DNA. Instead, it appears to bind to GyrA and interferes with gyrase action by preventing binding of the enzyme to DNA.[25] The *E. coli* protein GyrI, an inhibitor of gyrase, also appears to interfere with the binding of the enzyme to DNA; GyrI has been postulated to play a role in protecting cells against killing by gyrase poisons that trap the covalent gyrase–DNA complex.[26]

ESTABLISHING THE CELLULAR TARGET OF A DRUG

In the preceding section we saw how different classes of agents interact with the DNA topoisomerases. But these results by themselves are insufficient to establish the particular enzymes as the relevant cellular targets of the particular agents. This uncertainty arises from two potential complications. First, a drug often interacts with more than one cellular entity; that is, it may have multiple targets. Drugs like amsacrine and ellipticine, for examples, are DNA intercalators capable of inserting themselves between adjacent DNA base pairs (see Chapter 3), and hence can potentially interfere with any of a number of cellular processes that involve DNA. Anthracycline drugs, including doxorubicin and daunorubicin, are known to generate semiquinone free radicals as well as reactive oxygen species inside cells, and these reactive species can in turn damage various cellular components. The bisdioxopiperazines, originally developed to bind metal ions like Ca(II) and Fe(II), were long thought to manifest their physiological effects through metal-ion binding. The second complication in ascertaining particular enzymes as the relevant targets of specific agents is that physicochemical and structural studies are often done at drug concentrations that are rarely reached under physiological conditions. Therefore, the biological relevance of the resulting in vitro findings are sometimes in doubt.

---

25. Flatman, R.H., et al. 2005. *Antimicrob. Agents Chemother.* **49:** 1093–1100.

26. Nakanishi, A., et al. 1998. *J. Biol. Chem.* **273:** 1933–1938; Chatterji, M. and Nagaraja, V. 2002. *EMBO Rep.* **3:** 261–267; Chatterji, M., et al. 2003. *Arch. Microbiol.* **180:** 339–346.

Mutational analysis offers a powerful tool for establishing a cellular entity as the target of a particular agent. The initial clue that DNA gyrase is the cellular target of the plasmid-encoded toxin, for example, came from the mapping of CcdB resistance mutations to the *E. coli gyrA* gene. Similarly, mapping of coumermycin and nalidixate resistance to the *E. coli gyrB* and *gyrA* genes, respectively, was critical to identifying DNA gyrase as the target of these drugs.[5,7] Numerous mutational analyses of drug resistance or drug hypersensitivity, often in model systems ranging from phage T4-infected *E. coli* to mouse, have provided strong evidence linking DNA topoisomerases to the various compounds we have considered.

Despite possible complications (see shaded panel below), genetic approaches have been invaluable in identifying the cellular target of a drug. First, when combined with structural and biochemical studies, mutational information often goes far beyond target identification, and contributes significantly to obtaining a coherent picture of how a DNA topoisomerase-targeting drug exerts its physiological effects. Second, for some of the DNA topoisomerase-targeting drugs, identification of their targets can exploit what has been termed "sensitivity dominance."

Sensitivity dominance refers to a cell's being sensitive to a drug even when it bears two copies of the target gene of the drug, one encoding a protein that is sensitive and the other encoding a protein that is resistant to the drug. This dominance can be readily understood for drugs that trap the covalent DNA–enzyme complexes: In the case of the drug camptothecin, for example, the presence of camptothecin-resistant mammalian DNA topoisomerase I would have little effect on DNA damage mediated through the camptothecin-sensitive enzyme, and cells expressing both forms of the en-

Drug resistance does not necessarily involve mutations in the cellular target of a drug. For example, mutations that affect the import or efflux of a drug or drugs into cells can strongly affect drug sensitivity. In the case of drugs that inactivate a functionally indispensable enzyme, mutations in regulatory genes that lead to overexpression of the enzyme can make the cells more tolerant to these drugs. By contrast, for the "poison" class of drugs that convert a DNA topoisomerase to a DNA-damaging agent, an increase in the cellular level of the enzyme can produce a greater degree of DNA damage and hence increase drug sensitivity, whereas a decrease in the cellular level of the enzyme would result in drug resistance. The drug sensitivity of cells can also be affected by changes in gene products that interact with the primary targets of the drugs, one example being the *E. coli* protein GyrI described above. Furthermore, changes in resistance or sensitivity to a drug, especially to drugs that trap covalent DNA–DNA topoisomerase covalent complexes, can result from alterations in secondary or indirect processes or pathways involving the drug target. For example, mutations in genes involved in processing drug–enzyme–DNA ternary complexes, or in pathways connecting the ternary complexes to cell killing (including those involved in DNA damage response, repair, and damage-induced cell killing), can have strong effects on drug sensitivity.

zyme would therefore remain sensitive to the drug. Sensitivity dominance has been observed, for example, in yeast cells expressing two forms of human DNA topoisomerase IIα, one resistant and the other sensitive to the bisdioxopiperazine ICRF-193.[27] Here, yeast Top2 itself is also insensitive to ICRF-193, but locking a drug-sensitive human enzyme in its closed clamp conformation around the DNA is sufficient to cause cell death, probably by presenting a physical barrier to the normal progression of transcription assemblies along the DNA template.[27]

The results described for the two sensitivity dominance examples provide strong evidence that in each case the drug-sensitive enzyme is the primary target of the drug. There is also a special case in which the target DNA topoisomerase itself is functionally dispensable but can be converted to a DNA- damaging agent by a drug. In yeasts, for example, the *TOP1* gene encoding Top1 can be deleted without a significant effect on cell growth. It has been shown that yeast cells that are sensitive to camptothecin become resistant to the plant alkaloid upon deletion of the *TOP1* gene, and reintroducing a plasmid-borne yeast *TOP1* gene into the Δ*top1* cells restores camptothecin-sensitivity.[28] These results indicate that in yeast cells Top1 is the target of camptothecin, and the only significant target.

CELLULAR EVENTS FOLLOWING TERNARY COMPLEX FORMATION

Typically, the physiological effects of a drug that acts by inactivating a cellular entity can usually be understood in terms of the normal cellular functions of its target. In these cases, the effects of the drug are similar to those that result from reducing the cellular concentration of its target. This is not the case, however, for many of the DNA topoisomerase-targeting drugs: the antiproliferative and cytotoxic effects of these drugs may bear little relation to the cellular functions of the target enzymes. We have seen that deposition of tightly bound topoisomerase molecules on DNA can be detrimental to the cells because the tightly bound protein molecules may interfere with vital processes that occur along intracellular DNA. Among these processes, replication and transcription are two of the most significant. Thus, even a functionally dispensable DNA topoisomerase like yeast Top1 can be a very potent drug target. In the following sections, we shall first summarize the interactions between ternary complexes and replication and transcription, and then consider the cellular processing of these complexes.[29]

27. Jensen, L.H., et al. 2000. *J. Biol. Chem.* **275:** 2137–2146.
28. Nitiss, J. and Wang, J.C. 1988. *Proc. Natl. Acad. Sci.* **85:** 7501–7505; Bjornsti, M.A., et al. 1989. *Cancer Res.* **49:** 6318–6323.
29. For reviews, see Li, T.-K. and Liu, L.F. 2001. *Annu. Rev. Pharmcol.* **41:** 53–77; Pommier, Y. 2006. *Nature Rev. Cancer* **6:** 789–802.

Ternary Complex Formation and Replication

It is well established that cell killing by DNA topoisomerase poisons that trap covalent enzyme–DNA complexes is often closely tied to DNA replication. For example, long before the identification of Top1 as the target of camptothecin, the cytotoxicity of the plant alkaloid was known to be manifested mainly in the S phase of the cell cycle, during which the DNA replicates.[30] Following the finding that camptothecin stabilizes the covalent complex between DNA and its target enzyme, the interplay between an advancing replication fork and a camptothecin–Top1–DNA ternary complex was examined in SV40 replication systems in vitro and in vivo.[31] Prior to collision between a ternary complex and the replication machinery, the two ends of a severed DNA strand in a ternary complex are kept in close proximity by the topoisomerase, and resealing of the DNA strand can readily occur upon removal of the drug. In the event of a collision, however, several things may happen. First, the ternary complex, which is reversible prior to collision, may become irreversible, owing to a change in protein conformation resulting from the collision. In an irreversible ternary complex, the broken DNA strand can no longer be rejoined, even if the drug is removed. Second, when the elongating 3′ end of a replication fork approaches the DNA break in the ternary complex from the downstream side (Fig. 10-7), the shielded DNA strand break in the ternary complex may become exposed, thereby forming a bare or "frank" double-stranded DNA end, with the original 5′-OH group in the drug-trapped covalent complex at its 5′ end. The 5′ end of the broken DNA strand is no longer juxtaposed to the topoisomerase-attached 3′-phosphoryl group, and the two can no longer be rejoined by the attached topoisomerase whether or not the enzyme has been inactivated. Third, the roadblock may stall the replication fork, and the replication machinery may also suffer collateral damage in the collision, such as the loss of certain associated protein factors. The stalled replication fork may thus remain arrested at the collision site until restored to its functional form. Fourth, the replication machinery may force the reversal of the ternary complex and dissociation of the bound DNA topoisomerase. Even here the replication assembly may suffer damage and be in need of repair before it can resume DNA synthesis.

Conversion of a reversible ternary complex to an irreversible state has also been observed in a purified *E. coli* DNA replication system when *E. coli* Top4 is affixed covalently to the DNA template by norfloxacin.[32] Similarly, in several purified DNA replication systems, as well as in phage-infected *E. coli* cells, ternary complex formation involving bacterial and phage type II DNA topoisomerases has

30. Li, L.H., et al. 1972. *Cancer Res.* **32:** 2643–2650.

31. Avemann, K., et al. 1988. *Mol. Cell. Biol.* **8:** 3026–3034; Hsiang, Y.H., et al. 1989. *Cancer Res.* **49:** 5077–5082; D'Arpa, P., et al. 1990. *Cancer Res.* **50:** 6919–6924; Strumberg, D., et al. 2000. *Mol. Cell Biol.* **20:** 3977–3987.

32. Hiasa, H., et al. 1996. *J. Biol. Chem.* **271:** 26424–26429.

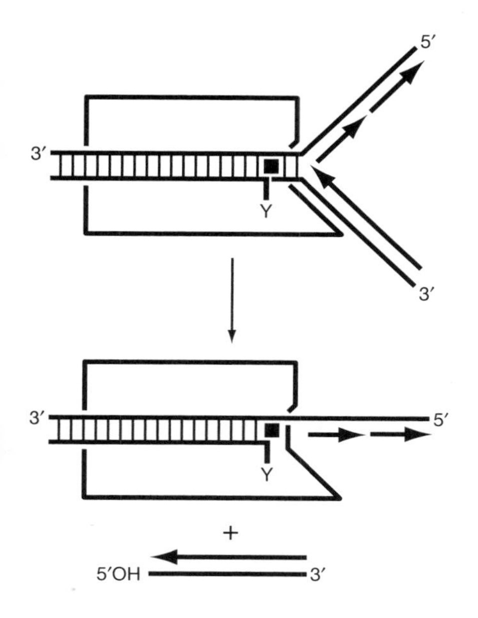

Figure 10-7. The collision between a replication fork and a camptothecin-trapped type IB DNA topoisomerase. The two parallel lines represent the unreplicated DNA double helix, and the vertical lines connecting them represent the base pairs between them. The arrows shown in the drawing on the *top* mark the directions of progeny strand extension, and the progression of the replication fork depicted is from right to left. The letter Y represents the active-site tyrosine of the type IB DNA topoisomerase (represented by the large box); it is covalently linked to the 3′ end of the −1 nucleotide owing to the insertion of a camptothecin (represented by a filled box) between base pairs −1 and +1. As the replication fork advances toward the trapped ternary complex from the direction shown in the figure, one arm of the fork may break off (*bottom* drawing). (Modified, with permission, from Hsiang, Y.H., et al. 1989. *Cancer Res.* **49:** 5077–5082.)

been shown to stop the advancement of a replication fork. In these cases, the arrest is sometimes accompanied by reversal of the ternary complex and dissociation of the topoisomerase from DNA.[33] In contrast to the case with the camptothecin-trapped covalent complex of Top1 and DNA, however, the formation of frank double-stranded DNA breaks is rarely seen in collisions between a replication fork and a drug-trapped DNA-linked complex involving a type II DNA topoisomerase. Rather, the exposed ends of the DNA in this latter case are generated by subsequent proteolytic and/or nucleolytic processing of the covalent complexes after a collision.[32,33]

The replication dependence of the cytotoxicity of some DNA topoisomerase-targeting drugs is closely tied to the dependence of their clinical efficacy on the scheduling of their administration. For the same cumulative dose of a replication-dependent "topoisomerase poison," for example, a few large doses may be less effective than continuous or frequent administration of low doses, because in the former case cytotoxicity would be limited to the population of tumor cells in the S phase of the cell cycle during the treatment windows.[34]

33. See, for example, Drlica, K. 1999. *Curr. Opin. Microbiol.* **2:** 504–508; Wentzell, L. M. and Maxwell A. 2000. *J. Mol. Biol.* **304:** 779–791; Hiasa, H. and Shea, M.E. 2000. *J. Biol. Chem.* **275:** 34780–34786; Hong, G. and Kreuzer, K.N. 2003. *Proc. Natl. Acad. Sci.* **100:** 5046–5051; Pohlhaus, J. and Kreuzer, K. 2005. *Mol. Microbiol.* **56:** 1416–1429.

34. See, for example, O'Leary, J. and Muggia, F.M. 1998. *Eur. J. Cancer* **34:** 1500–508; Houghton, P.J., et al. 1995. *Cancer Chemother. Pharmacol.* **36:** 393–403.

In the case of a type II enzyme, the fate of a drug–topoisomerase–DNA ternary complex during mitosis also poses an interesting question. In a ternary complex, the broken double-stranded DNA is held together by protein–protein interactions between the two halves of the enzyme. Thus when a duplicated pair of chromosomes is being pulled apart by microtubule filaments during mitosis, the protein–protein dimer interface in Top2 ternary complex may constitute a weak spot in the stretched DNA where chromosome breakage may occur. Experimental assessment of the strength of the Top2 dimer interface, and experiments probing the effects of mitotic chromosome segregation on Top2-ternary complexes, are needed to answer this question.

Ternary Complexes and Transcription

As described earlier, DNA topoisomerase II trapped on the DNA template by the bisdioxopiperazine class of inhibitors interferes with transcription because the trapped enzyme–inhibitor complex on the DNA template acts as a roadblock to the transcription machinery.[27] There is also strong evidence that drug-trapped covalent complexes of DNA with either Top1 or Top2 can block elongation of the transcribing polymerases.[29] In the case of the eukaryotic Top1 trapped on DNA by camptothecin, in vitro transcription studies using phage T7 RNA polymerase suggest that the effect is directional: Elongation by the phage enzyme is prevented, and the ternary complex converted to an irreversible state, only if the topoisomerase is covalently linked to the template strand of the DNA.[35]

Proteolytic Processing of DNA Topoisomerases Trapped on DNA

Inside cells, elaborate processes have evolved to deal with DNA topoisomerases and other proteins that become tightly bound to DNA, so as to restore normal transcription and replication-fork progression. If such attempts are overwhelmed, cell death often ensues. Hence the efficacy of DNA topoisomerase-targeting therapeutic agents is closely tied to the processing of DNA-trapped topoisomerases, to the repair of any damage to the DNA that may result from processing of the ternary complexes, and to the pathways that commit a cell to its own destruction (a process known as "apoptosis").

In mammalian cells, DNA topoisomerase I trapped on DNA by camptothecin is degraded by the 26S proteasome.[36] This proteolytic degradation is most likely unrelated to the presence of a broken DNA strand in the topoisomerase–DNA com-

35. Bendixen, C., et al. 1990. *Biochemistry* **29:** 5613–3619. Wu, J. and Liu, L.F. 1997. *Nucleic Acids Res.* **25:** 4181–4186.

36. Desai, S.D., et al. 1997. *J. Biol. Chem.* **272:** 24159–24164; Desai, S.D., et al. 2001. *Cancer Res.* **61:** 5926–5932; Desai, S.D., et al. 2003. *Mol. Cell Biol.* **23:** 2341–2350; Desai, S.D., et al. 2003. *Mol. Cell. Biol.* **23:** 2341–2350.

plex, because Top2β tightly locked around DNA by bisdioxopiperazines is also degraded by the same proteasome despite the apparent absence of a DNA break in the particular ternary complex.[37] The molecular details leading to the action of the 26S proteasome are largely unknown. It appears, however, that the action of the 26S proteasome is preceded by the covalent attachment of polypeptides of the ubiquitin family to the topoisomerase.[38] Conjugation of a number of these polypeptides to DNA topoisomerases that are covalently or noncovalently trapped on DNA has been observed, including conjugation of the 8.5-kDa ubiquitin monomer, polyubiquitins consisting of varying numbers of monomers, and ubiquitin-related polypeptides SUMO-1, SUMO-2 and SUMO-3. It is unclear how the extents of such modifications are regulated. Conjugation and deconjugation are reversible processes conducted by different groups of proteins, and the trapping of DNA topoisomerases on DNA appears to shift the equilibria in favor of conjugation.

Ubiquitination and polyubiquitination, but not sumoylation, have been shown to be closely related to the targeting of proteins for degradation. It is therefore likely that ubiquitination and polyubiquitination of trapped topoisomerases form an integral part of the proteasomal degradation of the trapped enzymes. The roles of sumoylation (SUMO-modification) of the trapped topoisomerases are less clear; sumoylation of a large number of proteins has been implicated in affecting a plethora of cellular processes, including DNA repair, regulation of transcription, changes in chromatin structure, and targeting of the modified proteins to various cellular compartments. Sumoylation of proteins is generally unrelated to channeling of the proteins for proteolysis. However, the finding that proteolysis of Top2β trapped on DNA by ICRF-193 is abolished upon inactivation of the SUMO-conjugating enzyme Ubc9 suggested that sumoylation of the drug–Top2β–DNA ternary complex is required for its degradation (see the paper by Isik et al. cited in Footnote 38). It is plausible, however, that the effect of Ubc9 on proteolysis of the trapped topoisomerase is mediated by modification of one or more participants in the proteolytic pathway and is not necessarily related to the direct modification of the trapped topoisomerase itself. In yeast, doxorubicin resistance is critically dependent on sumoylation, but this effect appears to be mediated through sumoylation of proteins controlling the intracellular concentration of doxorubicin, rather than through direct sumoylation of the DNA-linked topoisomerase.[39]

The second process that appears to be involved in the channeling of DNA-trapped topoisomerases to proteolytic degradation is transcription. Whereas ubiquitination and

37. Xiao, H., et al. 2003. *Proc. Natl. Acad. Sci.* **100:** 3239–2244.

38. Mao, Y., et al. 2000. *J. Biol. Chem.* **275:** 26066–26073; Fiorani, P. and Bjornsti, M.A. 2000. *Ann. N.Y. Acad. Sci.* **922:** 65–75; Mao, Y., et al. 2001. *J. Biol. Chem.* **276:** 40652–40658; Isik, S., et al. 2003. *FEBS Lett.* **546:** 374–378; Desai, S.D., et al. 2003. *Mol. Cell. Biol.* **23:** 2341–2350.

39. Huang, R.Y., et al. 2007. *Cancer Res.* **67:** 765–772.

sumoylation of trapped ternary complexes are independent of transcription, it appears that degradation of the trapped complexes by the 26S proteasome is transcription-dependent: Inhibition of transcription by the addition of transcription inhibitors, but not inhibition of replication, significantly reduces the 26S proteasome-mediated degradation of trapped topoisomerases.[37] Thus, the arrest of nascent transcripts by the drug–topoisomerase–DNA ternary complexes may be closely tied to the action of the 26S proteasome on these complexes.

Removal of DNA-Linked Peptides

Proteolysis of DNA topoisomerases that are covalently trapped on DNA is expected to leave short peptides at the DNA ends: 3′-phosphoryl ends in the case of a type IB enzyme, and 5′-phosphoryl ends in the case of a type II enzyme. One of the enzymes that might be involved in the removal of the attached peptides is a phosphodiesterase termed Tdp1. Tdp1 was initially thought to specifically hydrolyze phosphotyrosine bonds between a DNA 3′-phosphoryl group and a protein tyrosyl residue (hence the name tyrosyl phosphodiesterase 1).[40] Subsequently, however, Tdp1 was also found to hydrolyze 5′-phosphoryl-tyrosine,[41] phosphoamide links, and a variety of chemical moieties at 3′-phosphoryl ends of DNA.[42] In yeast, inactivation of Tdp1 leads to the hypersensitivity of cells to drugs that stabilize covalent complexes between DNA and Top1 and Top2, which is consistent with a role of Tdp1 in the removal of DNA topoisomerase remnants covalently linked to DNA ends. The phosphodiesterase may have a more general role, however, in the repair of single-stranded DNA breaks and the removal of protein-DNA crosslinks.[42,43]

Two additional classes of enzymes, acting either in conjunction with or in parallel to Tdp1, also appear to participate in processing DNA-linked DNA topoisomerases. Because Tdp1 leaves 3′-phosphoryl groups at the ends of DNA, one or more 3′-phosphatases would be required to convert these to 3′-hydroxyl groups to permit the extension of the DNA strands bearing them by DNA polymerases of the replication or repair systems.[43] Peptides covalently linked to DNA can also be excised by nucleases. Several nuclease complexes have been implicated in processing type I or II DNA topoisomerases that become covalently linked to DNA in the presence of drugs of the poison class. In the budding yeast, these include the Rad1–Rad10 dimer, the

40. Yang, S.W., et al. 1996. *Proc. Natl. Acad. Sci.* **93:** 11534–11539.

41. Nitiss, K.C., et al. 2006. *Proc. Natl. Acad. Sci.* **103:** 8953–8958.

42. Interthal, H., et al. 2005. *J. Biol. Chem.* **280:** 36518–36528.

43. For reviews, see Connelly, J.C. and Leach, D.R. 2004. *Mol. Cell* **13:** 307–316; Pommier, Y. 2006. *Nat. Rev. Cancer* **6:** 789–802; Pommier, Y., et al. 2006. *Prog. Nucleic Acid Res. Mol. Biol.* **81:** 179–229; el-Khamisy, S. F. and Caldecott, K.W. 2007. *Neuroscience* **145:** 1260–1266.

Mus81–Mms4 complex, the Mre11–Rad50–Xrs2 complex, and the *SLX1* and *SLX4* gene products.[43–45] The Mre11–Rad50-Xrs2 counterparts of *E. coli* and phage T4 have also been implicated in processing covalent DNA–protein complexes.[46] Interestingly, the nucleolytic action of Mre11–Rad50–Xrs2 is apparently involved in processing DNA-linked Spo11, a type II DNA topoisomerase-like protein that initiates meiotic recombination by cleaving double-stranded DNA and forming covalent protein–DNA adducts at the broken DNA ends (see Chapter 7).[47]

Cellular Responses to DNA Damage

As we have discussed, in the case of DNA topoisomerase poisons that trap the covalent enzyme–DNA intermediates, processing of the trapped adducts can form single-stranded gaps and/or frank double-stranded breaks in the DNA. Cells respond to such damage in diverse ways: Repair pathways are usually mobilized to repair DNA damage and restart arrested replication forks, and cell-cycle progression during the S and/or G2 phase is often delayed to give a cell more time to resume and complete DNA replication and to accomplish various repair tasks. When its repair efforts are overwhelmed or seem uneconomical, a cell may commit itself to apoptosis, an irreversible path leading to its death. These responses are clearly of key importance to the cytotoxic effects of the drugs, but they are also induced by DNA damage in general, such as damage caused by ionizing radiation or various chemical treatments. Therefore, the responses described here are a part of a cell's survival kit, and are not specific to topoisomerase-targeting drugs. Further discussion of these topics is beyond the scope of this book, and interested readers are referred to general texts on molecular biology and to treatises on DNA repair, cell-cycle regulation, and cell apoptosis.

THERAPEUTIC EFFICACY

For DNA topoisomerase-targeting drugs used in treating various microbial infections, like the quinolone class of antibiotics, it is clear that their specific action against a particular bacterium (but not mammalian cells) is closely tied to their specificity for particular bacterial enzymes. Ciprofloxacin, for example, effectively traps covalent complexes of DNA and bacterial type II DNA topoisomerases, but not those of DNA and mammalian type II DNA topoisomerases. Target specificity can also be explored

44. Vance, J.R. and Wilson, T.E. *Proc. Natl. Acad. Sci.* **99:** 3669–3674.

45. Liu, C., et al. 2002. *Proc. Natl. Acad. Sci.* **99:** 14970–14975; Deng, C., et al. 2005. *Genetics* **170:** 591–600; El-Khamisy, S.F., et al. 2007. *DNA Repair* **6:** 1485–1495.

46. Stohr, B.A. and Kreuzer, K.N. 2001. *Genetics* **158:** 19–28,; Connelly, J.C., et al. 2003. *DNA Repair* **2:** 795–807.

47. Neale, M.J., et al. 2005. *Nature* **436:** 1053–1057.

in the development of therapeutic substances against infective agents other than bacteria, including fungi, parasites, and viruses that encode their own DNA topoisomerases.

It is not apparent, however, why a DNA topoisomerase-active anticancer agent preferentially kills cancer cells relative to normal cells. Often, great heterogeneity is observed in the sensitivity of different or even the same types of tumors to a particular DNA topoisomerase-active anticancer agent. A comparison of six breast cancer epithelial cell lines, for example, showed a variation of several-hundred-fold in their camptothecin cytotoxicity.[48] In general, the efficacy of a drug depends on a complex set of differences between tumor and normal cells. The S-phase specificity of camptothecin derivatives, for example, makes these drugs more cytotoxic to proliferating tumor cells than to postmitotic normal cells. Indeed, the killing by these agents of normal proliferating cells, such as white blood cells and cells lining the gut, is a major factor contributing to their side effects (which include myelosuppression, or a decrease in blood cell production, and gastrointestinal disorders).

Tumor and normal cells may also differ in their cellular levels of the target DNA topoisomerases because of differences in the regulation of these enzymes at either the transcriptional or post-translational level in tumor and normal cells. Thus, for example, several tumor cell lines have been found to express a higher level of interferon gene stimulatory factor 15 (IGS15), a ubiquitin-like protein that is conjugated to many cellular proteins.[49] Extensive IGS15 conjugation in these lines appears to counter ubiquitin conjugation, leading to a reduction in protein degradation through the normal ubiquitin–26S proteasome pathway.[49] Consequently, the normal ubiquitin–26S proteasome-mediated decrease in Top1 or Top2 levels that follows the exposure of cells either to camptothecins or to bisdioxopiperazines is much less apparent in the tumor cell lines discussed here. This difference may in turn lead to greater intracellular concentrations of the target topoisomerase enzymes, and consequently to the enhanced sensitivity of tumor cells to the drug being used against them.[49]

The sensitivity of some tumor cells to DNA topoisomerase-targeting drugs of the poison class may also result from one or more defective DNA repair pathways in these cells, or from defective responses to DNA damage. Improper cell-cycle checkpoint control, for example, may result in defective repair of arrested replication forks and/or accumulation of single-stranded gaps or double-stranded breaks in intracellular DNA. In general, any difference between tumor and normal cells in pathways involved in the processing of drug–topoisomerase–DNA ternary complexes may significantly contribute to drug efficacy.

48. Davis, P.L. et al. 1998. *Anticancer Res.* **18:** 2919–2932.

49. Desai, S.D., et al. 2006. *Cancer Res.* **66:** 921–928.

DEVELOPMENT OF DRUG RESISTANCE[50]

The development of resistance to a previously effective drug is a major problem in clinical medicine. We will consider here three of the major mechanisms leading to drug resistance: (1) mutations in the target enzymes of drugs, (2) modulation of the intracellular concentrations of drugs by their membrane transporters, and (3) cellular activation and inactivation of drugs. Drug resistance owing to chemical modifications and/or metabolic changes of the drugs or their target enzymes is also discussed. Equally significant issues in the development of resistance to the DNA topoisomerase-targeting drugs include changes in DNA repair, cell-cycle regulation, and the control of cell death pathways; these topics are beyond the scope of this book, however, and we shall not consider them further.

Drug-Resistance Mutations in a Target DNA Topoisomerase

We have seen that mutations in a DNA topoisomerase can make cells refractive to drugs targeting the enzyme. For drugs that act by trapping covalent topoisomerase–DNA complexes, it is not necessary that the mutations occur within DNA sequences encoding the drug-binding pockets of the enzymes: any mutation that decreases topoisomerase activity or disfavors covalent enzyme–DNA complex formation can make such a drug less efficacious. Significantly, once a DNA topoisomerase acquires a mutation that reduces its enzyme activity or the efficiency with which it forms a covalent DNA–enzyme intermediate, cells expressing the mutant enzyme become more resistant to all drugs that act by trapping the DNA-linked enzyme. This type of resistance was first observed in a human leukemia cell line that became simultaneously resistant to etoposide, anthracyclines, mitoxantrone, and amsacrine, and was termed "atypical" multiple drug resistance so as to distinguish it from the "classical" multiple drug resistance caused by the acquisition of a more efficient drug efflux system (see below).[51]

Membrane Transporter Proteins and Drug Resistance

The ABC transporters are members of a large family of proteins (more than 50 in human cells) that get their name because all of them share a similar ATP-binding cassette.[52] These proteins catalyze the ATP-dependent transport of various substances across cell membranes, either into or out of cells, or between different cellular compartments. Individual transporters differ in their specificities, but a

50. Reviewed in Fojo, T. and Bates, S. 2003. *Oncogene* **22:** 7512–7523.

51. Danks, M.K., et al. 1987. *Cancer Res.* **47:** 1297–1301.

52. See Higgins, C.F. 2007. *Nature* **446:** 749–757; O'Connor, R. 2007. *Anticancer Res.* **27:** 1267–1272.

particular ABC transporter can often recognize a broad spectrum of structurally different substrates. The expression of several members of this family, including MDR1 (for multidrug resistance 1; also called ABCB1), MRP-1 (for multidrug resistance-related protein 1; also called ABCC1), and ABCG2, has been implicated in increasing the efflux of different therapeutic agents, reducing their cellular concentrations and hence increasing the resistance of cells to these drugs. Additionally, the non-ABC transporter RLIP76 has been shown to affect the cellular concentrations of multiple anticancer therapeutics.[53] The efficiency of a number of transporters appears to be affected by the cellular concentration of the cysteine-containing tripeptide glutathione (GSH). MRP-1 mediated resistance to doxorubicin in the human cell line MCF7, for example, has been reported to correlate with the cellular concentration of GSH.[54]

Drug and Enzyme Modifications

The most severe difficulty in the destruction of bacterial pathogens is their acquisition of drug resistance. This problem first appeared in the mid-1940s, shortly after the discovery of penicillin, and in the decades since then has become increasingly serious. As new and more potent antibiotics are developed, the bacteria fight back ferociously, often through the expression of genes that encode enzymes for modifying the antibiotics directly, or for modifying the cellular targets of the antibiotics. What has made the problem of drug-resistant bacteria even worse is that the drug-resistance determinants are often carried on plasmids that also carry genes for jumping in and out of the bacterial chromosome, as well as genes for transferring an integrated plasmid between the same or even different species. Recombination between these drug-resistance plasmids (sometimes called R factors) further enriches their weapon collection, and a single R factor may carry multiple drug-resistance determinants.

In the years following the introduction of the topoisomerase-targeting quinolones, drug resistance was caused primarily by mutations in the target DNA topoisomerases. More recently, however, two different types of plasmids carrying quinolone-resistance determinants have surfaced. One type encodes the protein Qnr, which, as is the case with the GyrI protein, interacts with DNA gyrase and protects it against the quinolones. The other type of plasmid expresses an acetylation enzyme that has been found to reduce the effectiveness of ciprofloxacin by placing an acetyl group on an amino nitrogen of the piperazine ring of this drug.[55]

In the case of anticancer drugs, cellular metabolism often critically affects their therapeutic efficacy. The Top1-targeting drug irinotecan, for example, is largely an

53. Awasthi, Y.C., et al. 2007. *Curr. Drug Metab.* **8:** 315–323.

54. Benderra, Z., et al. 2000. *Eur. J. Cancer* **36:** 428–434.

55. For a review, see Robicsek, A, et al. 2006. *Lancet Infect. Dis.* **6:** 629–640.

inactive "prodrug," and must be activated by cellular carboxylesterases to its active metabolite SN-38. Two metabolic pathways then process SN-38 for excretion; these pathways involve enzymes like UDP-glucuronosyltransferase 1A1 (UGT1A1), which converts SN-38 to a β-glucuronide, and cytochrome P450 3A4, which converts SN-38 into hydroxylated derivatives.[56] Significantly, the genes encoding UGT1A1 and P4503A4 are both polymorphic, and individual patients may carry different alleles of these genes. Thus, in addition to affecting irinotecan efficacy, these genes may account for variations in the severity of side effects of the drug in different patients.

The P450 family of genes is particularly important in drug activation and inactivation. Some 60 members of this family are known; they are highly polymorphic, and hundreds of different alleles have been identified. Besides its importance in the bioconversion of SN-38, the P4503A4 isozyme has been implicated in oxidation of the anticancer drugs etoposide and doxorubicin in the treatment of osteosarcoma.[56] Consequently, it is plausible that the selective expression of certain P450 isozymes in different tumors may contribute to drug resistance.

CANCER CHEMOTHERAPY AND CARCINOGENESIS

The cytotoxicity of all topoisomerase-targeting therapeutic agents now in clinical use can be attributed to DNA damage and/or interference with normal cellular transactions of DNA. Their effects on DNA, mediated through the DNA topoisomerases, make them potent drugs. Yet wielding a sharp knife at the genetic material carries substantial intrinsic risks. A case in point is the treatment-related development of secondary malignancies.

There is strong evidence that prior chemotherapy with the Top2-targeting drugs etoposide and anthracyclines greatly increases the probability of development of acute myeloid leukemia (AML) and myelodysplasia (MDS), the latter of which is sometimes termed preleukemia because it often progresses to AML.[57] The development of these treatment-related secondary malignancies, which have been termed t-MDS and t-AML and which accounts for 10%–20% of all known cases of MDS and AML, does not occur only in patients with prior exposure to topoisomerase-targeting drugs; exposure to radiation and alkylating agents that induce lesions in DNA also greatly increases the risk of subsequent development of these secondary malignancies. Patients with t-MDS and t-AML display chromosome abnormalities, including chromosome gain or

---

56. See, for example, Yanase, K. and Andoh, T. 2004. Cellular resistance to DNA topoisomerase I-targeting drugs. In *DNA topoisomerases in cancer chemotherapy* (ed. T. Andoh), pp. 129–143. Kluwer Academic/Plenum Publishers, New York; Dhaini, H.R., et al. 2003. *J. Clin. Oncol.* **21:** 2481–2485; Rochat, B. 2005. *Clin. Pharmacol.* **44:** 349–366; Rodriguez-Antona, C. and Ingelman-Sundberg, M. 2006. *Oncogene* **25:** 1679–1691.

57. For a review, see Pedersen-Bjergaard, J., et al. 2002. *Blood* **99:** 1909–1912.

loss, fragmentation, or balanced translocations with little net gain or loss of chromosome material.[58] These observed chromosome aberrations are consistent with the notion that the underlying cause of these secondary malignancies is tied to the inherent risk of treatments based on damaging DNA.

Significantly, with regard to secondary malignancies associated with etoposide, studies using mutant mice lacking Top2β in their skin cells indicate that carcinogenesis leading to melanoma of the skin is mediated mainly through the Top2β isozyme.[59] Because killing of cancer cells by etoposide is mediated mainly through the Top2α isozyme, there appears to be a major difference between the roles of the α and β isozymes in cancer treatment and cancer induction. This difference is probably related to the association of the Top2α isozyme with proliferating cells and the Top2β isozyme with postmitotic cells, but the precise underlying mechanism remains unknown.

WHY SO MANY TOPOISOMERASE-TARGETING DRUGS?

At first glance it would seem surprising that so many of the therapeutic agents in clinical use target the DNA topoisomerases. But it is no surprise that the central role of the DNA within cells makes it a prime target for cellular warfare. The DNA topoisomerases, with their multiple tasks, all of which require breakage and rejoining of one or both strands of the DNA double helix, have provided a convenient beachhead for assaults against DNA. The structural and mechanistic richness of the DNA topoisomerases has provided multiple platforms both for the evolution of natural compounds targeting these enzymes and for the development of new topoisomerase-active drugs in the laboratory.

Whereas drugs targeting the DNA topoisomerases are well represented in battles against external invaders such as infectious agents and internal rebels such as various cancers, there remains ample opportunity for expanding the current repertoire of topoisomerase-targeting therapeutics. Several classes of such compounds, such as the camptothecin derivatives, were discovered or introduced into clinical use only recently, and the further development of similar agents is likely to be fruitful. Even for an extensively studied class of drugs like the quinolone antibiotics, drug development remains a continuing struggle in light of the ever more serious problem of drug resistance, which often follows the introduction of a new drug. It has been suggested, for example, that the development of dual targeting quinolones that are equally effective against both DNA gyrase and Top4 may significantly reduce the likelihood of development of drug resistance, because a bacterium would have

58. For a summary based on examination of the chromosomes of some 500 patients, see Slovak, M.L., et al. 2002. *Genes Chromosomes Cancer* **33:** 379–394.

59. Azarova, A.M., et al. 2007. *Proc. Natl. Acad. Sci.* **104:** 11014–11019.

to concomitantly acquire mutations in both target enzymes to become resistant to these drugs.[60]

It is also curious that no effective drug has yet emerged against the type IA DNA topoisomerases. Because of the mechanistic similarity of the type IA and type IIA DNA topoisomerases in their breakage and rejoining of DNA strands, agents that interfere with type IA-catalyzed DNA breakage and rejoining probably already exist among the large number of compounds that have been collected as potential drugs for use against type IIA topoisomerases. Thus, further efforts in developing agents targeting the type IA enzymes are likely to be fruitful.[61] Inactivation of both type IA DNA topoisomerases is lethal, which should make it feasible to develop type IA topoisomerase-targeting compounds as antimicrobial therapeutics that act either as enzyme inhibitors or as poisons.[60] At present, there is insufficient information about the potential of the type IA enzymes as anticancer-drug targets. These enzymes are essential in the proper resolution of certain recombination intermediates, but it is unclear whether there are differences in the processing of such intermediates in cancer and in normal cells, and how such differences might be exploited in drug development.

The finding of poxvirus-like type IB topoisomerases in many bacteria, including the human pathogens *Mycobacterium avium* and *Pseudomonas aeruginosa*, has also stimulated interest in developing special type IB enzyme-targeting drugs for treating infections caused by these pathogens as well as poxvirus infections.[62] Additionally, because of the pivotal role of mitochondria in energy metabolism as well as in the commitment of cells to programmed cell death, cell-specific targeting of mitochondria has received substantial interest in cancer chemotherapy.[63] Thus, the discovery of a mitochondrial type IB DNA topoisomerase may provide a potential new target in the development of drugs of this class.[64] In view of the prominent roles of the DNA topoisomerases in Nature's battlefields, more therapeutics targeting this family of enzymes are most likely to emerge in the years to come.

60. Zhao, X., et al. 1997. *Proc. Natl. Acad. Sci.* **94:** 13991–13996; Pan, X.-S. and Fisher, L. M. *Antimicrob. Agents Chemother.* 1998. **42:** 2810–2816.

61. Cheng, B., et al. 2007. *J. Antimicrob. Chemother.* **59:** 640–645; Tse-Dinh, Y.C. 2007. *Infect. Disord. Drug Targets* **7:** 3–9.

62. Krogh, B.O. and Shuman, S. 2002. *Proc. Natl. Acad. Sci.* **99:** 1853–1858.

63. Armstrong, J.S. 2007. *Br. J. Pharmacol.* **151:** 1154–1165; Scatena, R., et al. 2007. *Am. J. Physiol. Cell. Physiol.* **293:** C12–21.

64. Zhang, H., et al. 2001. *Proc. Natl. Acad. Sci.* **98:** 10608–110613.

APPENDIX I

Identification of the Covalent Intermediates in Topoisomerase-Catalyzed Reactions

One hallmark of reactions catalyzed by all DNA topoisomerases is the formation of an enzyme–DNA covalent intermediate in which a DNA phosphorous atom is linked to the oxygen atom of an enzyme tyrosyl residue. Some of the experiments confirming its formation, and evidence that confirms it as the true intermediate in the topoisomerase-catalyzed DNA breakage and rejoining, have already been summarized in Chapter 2.

We now describe several key experiments that established the chemical identity of these enzyme-DNA covalent complexes. These experiments illustrate the interplay between hypotheses and experimental tests, and the cumulative nature of scientific knowledge—what is already known often serves as the base for adding a new piece to a puzzle. In addition to establishing the transesterification mechanism postulated for topoisomerase-catalyzed DNA breakage and rejoining, the identification and characterization of the covalent intermediates are important in two respects. First, DNA topoisomerase-targeting antimicrobial and antitumor drugs presently in clinical use all work by interacting with these covalent intermediates. Second, the formation of such covalent intermediates may also expose intracellular DNA to additional risks of chemical damage, because the presence of strand breaks, even transient ones, introduces weak spots in the DNA strands. Thus, similar to the topoisomerase-targeting drugs that act on these covalent intermediates, chemical carcinogens might also act at these sites with devastating consequences. Both of these aspects are discussed in Chapter 10.

Historically, elucidation of the chemical nature of the enzyme–DNA covalent complexes proceeded in three stages. First, the protein-free DNA end formed when a DNA topoisomerase becomes covalently linked to a DNA strand was identified, which in turn provided information on the enzyme–DNA covalent link itself. For example, if the protein-free end is a 3′-hydroxyl group, then the protein is most likely covalently linked to a 5′-phosphoryl end of the broken DNA strand (see Fig. 2-4). Similarly, if the protein-free end is a 5′-hydroxyl group, then the protein is most likely linked to a DNA 3′-phosphoryl end.

The simplest approach to identifying a chemical group at the end of a DNA strand was to take advantage of the known specificities of enzymes that act on DNA ends. The addition of a protein-denaturant to trap the covalent intermediate between a DNA topoisomerase and DNA (Chapter 2) also exposes the protein-free end of the broken DNA strand. When alkali was used to trap a covalent complex, then the pH and ionic makeup of the reaction mixture could be subsequently adjusted, and an enzyme that acts at the ends of DNA strands with known specificities could be used to determine the chemical nature of the protein-free end. An enzyme called polynucleotide kinase, for example, can transfer the terminal or γ-phosphate of ATP to a 5′- but not a 3′-hydroxyl end, or, in the case of a 5′-phosphoryl group at the end of a DNA strand, catalyze the exchange between the DNA 5′-phosphoryl group and the ATP γ-phosphate. Thus by using ATP with a ^{32}P- or ^{33}P-radiolabeled γ-phosphate (the ordinary phosphorous is ^{31}P, which is nonradioactive), a DNA with a 5′-hydroxyl or phosphoryl end could be tagged with a radioactive phosphate group at its 5′-end by the use of polynucleotide kinase, but a DNA with a 3′-end would remain unlabeled under the same reaction conditions. Another enzyme, called terminal transferase (or, more specifically, terminal deoxynucleotidyl transferase), can add nucleotides to a DNA 3′- but not to a DNA 5′-hydroxyl end in the presence of an appropriate nucleoside triphosphate. Among the DNA exonucleases (Chapter 1), those like *Escherichia coli* exonuclease I and the exonuclease activity of phage T4 DNA polymerase can nibble a DNA strand from its 3′-hydroxyl end (Fig. A1-1, I and II), and by contrast phage λ exonuclease nibbles a DNA strand from a 5′-hydroxyl end (Fig. A1-1, III).

Through the use of the termini-specific enzymes described above, it was shown that the type IB DNA topoisomerases form covalent intermediates with a 5′-hydroxyl group at the protein-free end of the broken DNA strand, and all the other DNA topoisomerases form covalent intermediates with a 3′-hydroxyl group at the protein-free end of the broken DNA strand.[1–3] These results in turn suggested that an enzyme moiety is joined to a DNA 3′-phosphoryl group in the covalent intermediate between DNA and a type IB enzyme, and in all other cases a topoisomerase residue is joined to a DNA 5′-phosphoryl group.[1–3]

1. Champoux, J.J. 1977. *Proc. Natl. Acad. Sci.* **74:** 3800–3804; 1978. *J. Mol. Biol.* **118:** 441–446.
2. Sugino, A., et al. 1977. *Proc. Natl. Acad. Sci.* **74:** 4767–4771; Gellert, M., et al. 1977. *Proc. Natl. Acad. Sci.* **74:** 4772–4776.
3. Depew, R.E., et al. 1976. *Fed. Proc.* **35:** 1493; 1978. *J. Biol. Chem.* **253:** 511–518.

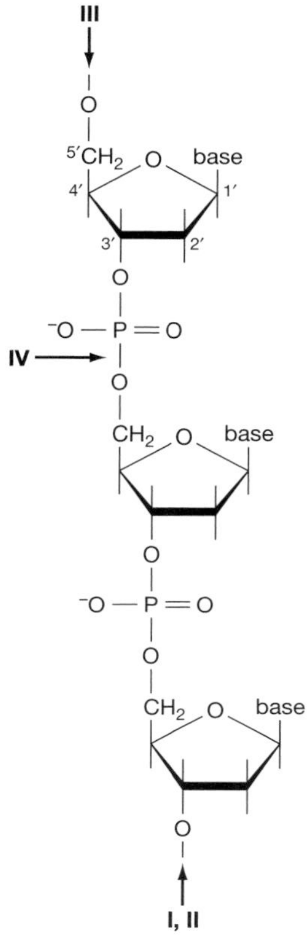

Figure A1-1. Sites of action of some of the nucleases used in determining the chemical bonds joining DNA topoisomerases to DNA in the enzyme-protein covalent intermediates. Only three nucleotides are shown in the depicted DNA strand. *E. coli* exonuclease I (I) and the exonuclease activity of phage T4 DNA polymerase (II) nibble a DNA strand from a 3′-hydroxyl end, and a phage λ exonuclease nibbles a DNA strand from a 5′-hydroxyl end (III). *Staphylococcus* nuclease (IV) hydrolyzes a phosphodiester bond between the phosphorous atom and the oxygen atom attached to the 5′C to form a 5′-phosphoryl and 3′-hydroxyl group at the ends of the severed DNA strand.

In the second stage, the enzyme moiety that becomes linked to a DNA strand during the formation of the covalent intermediate was identified. The stability of a particular DNA–enzyme covalent complex to treatment with acid, base, other chemicals, or high temperature, etc. was first tested, and these results provided information on the likely candidates for the covalent link. Further experiments were then designed to test these initial guesses. In the case of *E. coli* Top1, for example, the chemical stability of the covalent intermediate suggested that the covalent link might be a phosphotyrosine bond. To test this possibility further, radiolabeling of the phosphoryl groups of a DNA with ^{32}P was first performed by culturing bacteria in a synthetic medium containing labeled orthophosphate, and the ^{32}P-labeled DNA was then used in covalent complex formation with *E. coli* Top1. As it was known that the protein in the covalent intermediate would mosty likely be linked to a DNA 5′-phosphoryl group, the covalent complex trapped by alkali addition was digested with staphylococcal nuclease (also called micrococcal nuclease) after adjusting the pH of the reaction medium. This nuclease, purified from the bacterium *Staphylococcus aureus*, was chosen because it cleaves

DNA (or RNA) internucleotide P—O bonds at their 5′ position, and thus it would be expected to leave a DNA phosphoryl group on the protein after it removes much of the enzyme-linked DNA strand (see position indicated by IV in Fig. A1-1). Some ^{32}P was indeed found to be associated with *E. coli* Top1 after nuclease treatment of its covalent complex with DNA. The ^{32}P-label protein was then subject to prolonged treatment with dilute hydrochloric acid to break the peptide bonds between amino acid residues in the protein. A large fraction of the protein–phosphate link also suffered acid hydrolysis during this harsh treatment, but a minor fraction survived to yield an amino acid with a radiolabeled phosphorus attached to it. This product was identified to be phosphotyrosine by comparing its properties with those of chemically synthesized phosphotyrosine.[4] Similar experiments with mammalian Top1, using nucleases other than the staphylococcal enzyme, also showed that in its covalent intermediate with DNA the oxygen atom in a tyrosyl residue is linked to a DNA 3′-phosphoryl group.[5]

In the third and final stage, the locations of the particular tyrosyl residues involved in covalent complex formation between DNA and various DNA topoisomerases were determined. In a method reported in 1987, the covalent complex between DNA and a particular DNA topoisomerase was first isolated. The protein was then unfolded and exhaustively digested with a protease that specifically cleaves at particular amino acid residues when the protein is denatured, that is, in the form of an unfolded chain. Such treatments would thus leave a short peptide on the DNA, and this DNA-attached peptide could be separated from all other peptides in the digest through purification of the DNA attached to it. The product of this final purification was then loaded in an amino acid sequenator, an automated protein-sequencing instrument. In such an instrument, amino acid residues from the amino-terminus of a peptide are sequentially released by chemical treatment and analyzed, one per reaction cycle. For the *E. coli* gyrase-DNA covalent intermediate, the DNA-linked peptide after extensive digestion with an enzyme trypsin, which cleaves a polypeptide chain at the carboxyl side of lysyl or arginyl residues, was purified and analyzed in this way. No amino acid was detected in the first cycle, and the next four cycles gave the sequence Thr-Glu-Ile-Arg. Searching the amino acid sequence of GyrA showed a perfect match from residue 123 to 126, which is preceded by a tyrosyl residue at position 122 and an arginyl residue at position 121; no other match of the quartet was found. This result shows that Tyr-122 of *E. coli* GyrA is the site of DNA attachment: Cleavage by trypsin on the carboxyl side of the lysinyl and argininyl residues would yield a DNA-attached peptide Tyr122-Thr123-Glu124-Ile125-Arg125; automated sequencing of this peptide would show a blank in the first cycle, because the covalent attachment of Tyr-122 to DNA prevented its release as an amino acid.[6] Applying the same method to the co-

4. Tse, Y.-C., et al. 1980. *J. Biol. Chem.* **255:** 5560–5565.

5. Champoux, J.J. 1981. *J. Biol. Chem.* **256:** 4805–4809.

6. Horowitz, D.S. and Wang, J.C. 1987. *J. Biol. Chem.* **262:** 5339–5344.

valent intermediates of *E. coli* and *S. cerevisiae* DNA topoisomerase I then identified Tyr-319 and Tyr-727 of the respective enzymes as the sites of covalent attachment to DNA.[7]

As a final confirmation, site-directed mutagenesis was performed to replace the nucleotides encoding the particular tyrosyl residues, namely Tyr-122 in *E. coli* GyrA subunit, Tyr-727 in *S. cerevisiae* DNA topoisomerase I, and Tyr-319 in *E. coli* DNA topoisomerase I, by those encoding a phenylalanine, serine, or alanine, and in all cases the mutant enzymes were found to be inactive.[6,7] Thus the particular tyrosyl residue identified in each case is the one positioned at the catalytic site of the enzyme, and its attack of a DNA phosphorous leads to breakage of the DNA strand. Finally, because all members of the same subfamily of DNA topoisomerases share extensive similarities in their amino acid sequences, once the active-site tyrosine of a representative member of a subfamily is experimentally determined, the active tyrosines of all other members of the same subfamily can be deduced from the amino acid sequences of the enzymes.[6,7]

7. Lynn, R.M. and Wang, J.C. 1989. *Proteins* **6:** 231–239; Lynn, R.M., et al. 1989. *Proc. Natl. Acad. Sci.* **86:** 3559–3563.

APPENDIX 2

Catalysis of DNA Breakage and Rejoining by DNA Topoisomerases

ENZYMES ARE NATURE'S WONDERS THAT HAVE EVOLVED over billions of years to sustain all life-forms on Earth. For the breakage and rejoining of a DNA strand by a DNA topoisomerase, the enzyme and the DNA strand must first be precisely positioned, relative to each other, so that the key chemical groups in both can assume their proper locations for transesterification to occur. Intricate interactions between the two macromolecular partners require the participation of many amino acid side chains as well as some of backbone groups of the enzyme, and the precise positioning of these groups in turn demands that the polypeptide chain forming the enzyme be properly folded into a well-configured three-dimensional structure. On the DNA side, often the base and backbone groups are both involved, and the scissile phosphorous assumes its central role in the transesterification reactions.

The binding of a DNA topoisomerase to DNA is often accompanied by substantial structural adjustments in one or both macromolecules. For example, a short stretch of base pairs of the DNA is unpaired during the formation of a type IA enzyme–DNA complex (Chapter 3), and the binding of the type IB vaccinia topoisomerase to a specific DNA sequence appears to elicit structural alterations to position the active-site residues for attacking the scissile phosphorous.[1]

Chemically, transesterification between DNA and DNA topoisomerase goes through a transient intermediate in which the DNA phosphorous is surrounded by five oxygen atoms, four of which are those originally associated with the scissile phosphorous, and the fifth comes from the hydroxyl group of the active-site tyrosyl residue of the enzyme (see Chapter 4 and Fig. 4-1b; this hydroxyl group is some-

1. Cheng, C., et al. 1998. *Cell* **92:** 841–850. The structure of a closely related enzyme of the bacterium *Deinococcus radiodurans* shows, however, that in this enzyme the catalytically important side chains are all properly positioned even in the absence of bound DNA; see Patel, A., et al. 2006. *J. Biol. Chem.* **281:** 6030–6037.

times referred to as the "phenolic hydroxyl" because it resembles the same group in phenol). These five O atoms form a bipyramid around the central phosphorous, with the tyrosyl oxygen (O4) and a bridging oxygen in the DNA strand facing each other from opposite apices of the bipyramid (see Fig. 4-1b; the bridging oxygen in the DNA backbone is an O5′ in the case of a type IB enzyme, and an O3′ in the other cases).

In the absence of the intricate interactions between the enzyme and DNA strand, the bipyramid structure described above would be in a much higher energy state than a pair of DNA phosphoryl and tyrosyl hydroxyl groups, and the chance of its formation would be extremely low. Indeed, when DNA strands are simply mixed with the amino acid tyrosine itself, or with simple derivatives of tyrosine, transesterification between DNA and tyrosine is undetectable under physiological conditions. Thus a DNA topoisomerase must accomplish several remarkable feats to catalyze the reaction. Interactions between the bipyramid structure and specific residues of a DNA topoisomerase are used to stabilize the bipyramid structure and facilitate its formation. The bipyramidal transient intermediate is an example of a "transition-state complex" in a reaction pathway, and stabilization of a transition-state complex is often of key importance in enzymological as well as chemical catalysis.[2]

Factors other than transition-state stabilization are also significant in transesterification between DNA and DNA topoisomerase.[2] In Figure 4-1b the drawing for the reaction of the type IA enzyme, only the O4 atom of the active-site tyrosyl residue and the O3′ atom of the deoxyribose are shown in the reaction step, and the hydrogen atoms attached to O4 and O3′ were omitted for clarity. In the actual reaction, a proton (a hydrogen atom with a positive charge) must be removed from the tyrosyl hydroxyl, and added to the ribosyl O3′ of the DNA backbone to give a protein-linked 5′-phosphoryl and a protein-free 3′-hydroxyl end of the transiently broken DNA strand. Conversely, in the rejoining of the DNA strand, a proton must be stripped from the 3′-hydroxyl at the protein-free end, and added to the tyrosyl O4. The situations with the other subfamilies of DNA topoisomerases are basically the same, except that an O5′ rather than O3′ is involved in the case of the IB subfamily (see Fig. 4-1).

Ordinarily, at a neutral pH the hydroxyl group of a protein tyrosyl side chain does not like to part with its O4-attached proton, and dissociation of this proton becomes favorable only when the pH of the solution is above 9. Thus it has often been postulated that a certain group –B: with a pair of unshared electrons may be interacting with this proton, to help its removal from the tyrosyl O4 oxygen in the DNA cleavage reaction. The group –B: is called a "general base" because it acts as the acceptor of a proton, an acid. In the DNA rejoining reaction, the ribosyl hydroxyl is even less in-

2. For discussions on transition-state stabilization and other important features in enzyme catalysis, including general acid–base catalysis and electrostatic effects, see Fersht, A. 1985. *Enzyme structure and mechanism*, 2nd ed. Freeman, New York; 1998. *Structure and mechanism in protein science: A Guide to enzyme catalysis and protein folding*. Freeman, New York.

clined to let go of its proton, and a general base would seem particularly desirable to make the departure of the proton less painful.

The complement of a general base is a "general acid," which is any chemical group that readily donates a proton. In the reaction depicted, a general acid could assist the reaction by donating a proton to the departing ribosyl group; in the DNA rejoining reaction, a general acid could aid the departure of the DNA-linked tyrosyl O4. In enzyme catalysis, a general base and a general acid often go hand-in-hand; if a group –B: serves as a general base, then after picking up a proton the protonated form –B:H^+ would behave as a general acid in the reverse step. Mechanisms of enzyme-catalyzed reactions that involve general acid and general base groups are termed "general acid–base catalysis." The presence of an oppositely charged group in the vicinity of a departing charged group could also make the dissociation less costly energy-wise. In the case of a departing proton, for example, the dissociation reaction would be favored by the presence of a neighboring positively charged group. This type of mechanism is termed "electrostatic catalysis."[2]

The transesterification reactions catalyzed by various DNA topoisomerases share many common mechanistic features, especially among the type IA, IIA, and IIB enzymes, all of which possess a cluster of acidic residues that appear to be catalytically important, as well as a highly conserved arginyl residue near the active-site tyrosine (Chapters 3 and 5). It is the DNA breakage and rejoining reactions by the type IB enzymes, however, that have been studied most extensively.[3] The solution of a number of crystal structures of the enzyme–DNA complexes in the reaction path has greatly helped the elucidation of the mechanistic details of the type IB enzyme catalyzed reactions. The structures of the noncovalent and covalent complexes described in Chapter 4 have provided a great deal of information about the molecular participants before and after the DNA breakage reaction, and a structure that mimics the transition state of the reaction has also been obtained.[4] In this structure, the enzyme was that of a parasite *Leishmania donovani* (see Chapter 9), and the DNA the same 22-bp DNA used in the human Top1 cocrystal structures (Chapter 4). In this case, however, a nick with a 5′-OH on one side and a 3′-OH on the other side was placed in between the –1 and the +1 nucleotide of the scissile strand. The presence of this nick created a space to accommodate a V^{+5} vanadate ion to this site in the crystal structure of the DNA-enzyme complex. The vanadate ion is often used to mimic a transition-state penta-coordinated phosphorus, because it can readily interact with five oxygens to form a stable bipyramid structure with the metal ion at its center. In the illustration shown in Figure A2-1, the central V^{+5} is surrounded by five oxygens in a slightly distorted bipyramidal geometry, with its apical positions occupied by the phenolic oxygen of the active-tyrosyl residue Tyr222 (which corresponds to Tyr723

3. Reviewed in Champoux, J.J. 2001. *Annu. Rev. Biochem*. **70**: 369–413; Shuman, S. 1998. *Biochim. Biophys. Acta*. **1400:** 321–337. See also Krogh, B.O. and Shuman, S. 2000. *Mol. Cell* **5:** 1035–1041; Tian, L., et al. 2005. *Structure* **13:** 513–520; and Yakovleva, L., et al. 2008. *J. Biol. Chem*. **283:** 16093–16103 and references therein.

4. Davies, D.R., et al. 2006. *Mol. Biol.* **357:** 1202–1210.

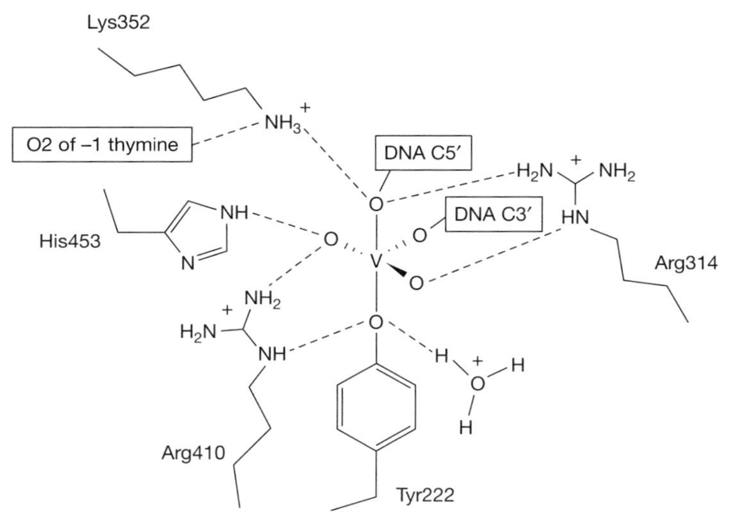

Figure A2-1. Chemical groups surrounding the bipyramidal vanadate in a DNA-type IB DNA topoisomerase-vanadate ternary complex. This ternary complex serves as a model for the transition-state complex in DNA breakage and rejoining by a type IB enzyme. (Adapted, with permission of Elsevier, from Davies, D.R., et al. 2006. *J. Mol. Biol.* **357:**1202–1210.)

in human Top1 or hTop1 and Tyr274 in vaccinia topoisomerase or vTop) and the DNA O5′ atom at the pre-existing nick. The apical oxygens in the depiction correspond to those of the attacking and leaving groups in the normal DNA–enzyme transesterification reactions. Together, O5′ and the three equatorial oxygens, including the O3′ atom, constitute the equivalent of the four oxygens of the scissile phosphate group in the DNA strand that would remain covalently linked to the C3′ of the cleaved DNA strand. In addition to the active-site tyrosine, the side-chains of the four highly conserved amino acid residues of the type IB enzymes, Arg314, Lys352, Arg410, and His453 (which correspond respectively to Arg488, Lys532, Arg590, and His632 of hTop1 and Arg130, Lys167, Arg223, and His265 of vTop), are all within hydrogen-bonding distances to one or more of the five oxygens around the pentavalent vanadium ion. In addition, Lys352 is also within hydrogen-bonding distance to O2 of the –1 base of the scissile strand, and a solvent water also appears to be H-bonded to O4 of the active-site tyrosine.

The *Leishmania* crystal structure and other structures described earlier suggest that transition-state stabilization, electrostatic effects and general acid–base catalysis are probably all significant in DNA breakage and rejoining by a DNA topoisomerase. The structure of the *Leishmania* type IB enzyme-DNA-vanadate ternary complex suggests, for example, that Arg410, or the solvent water interacting with the tyrosyl phenolic oxygen, might play the role of a general base and Lys352 and/or Arg314 the role of a general acid in DNA cleavage; that Arg410 might exert significant electrostatic effect in stabilizing the phenolic anion during the reaction; and that several of the residues

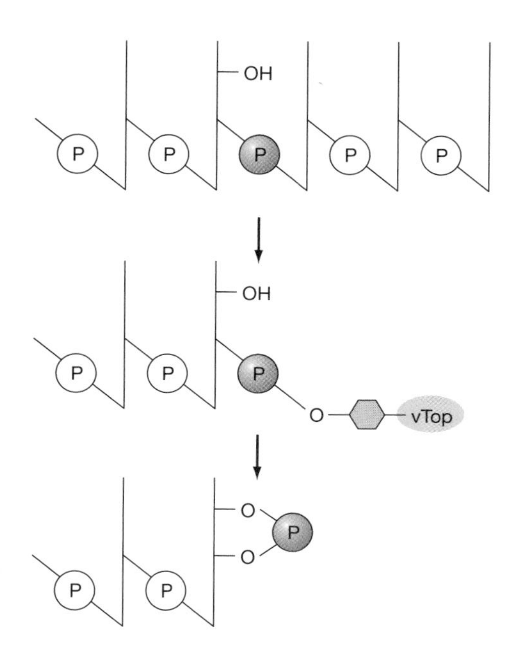

Figure A2-2. Cleavage at a ribonucleotide by vaccinia topoisomerase. The scissile strand contains a ribonucleotide at position –1, and strand breakage forms a covalent intermediate in which the enzyme (shaded oval labeled vTop in the figure) is linked to the ribosyl O3′. Attack of the enzyme-linked phosphorous by the vicinal 2′-hydroxyl group of the ribose forms a 2′, 3′-cyclic phosphate and breaks the protein–nucleic acid covalent intermediate.

surrounding vanadate might stabilize the bipyramidal reaction intermediate in the normal transesterification reactions.[4]

A DNA topoisomerase must provide an ideal environment for the active-site tyrosine hydroxyl to initiate DNA breakage, but also it must minimize the chance of this tyrosyl hydroxyl being usurped by a chemically similar group. If the strategically located tyrosyl O4 is replaced by a water oxygen atom, then DNA strand breakage by the enzyme would form a phosphoryl group at the broken DNA end that would normally become covalently joined to the enzyme tyrosyl; the DNA topoisomerase would then act as an endonuclease that hydrolyzes DNA backbone bonds. It has been observed, for example, that if a glutamate or histidine replaces the active-site tyrosine of vaccinia topoisomerase, then the topoisomerase becomes an endonuclease.[5] Similarly, in the DNA rejoining reaction a properly located hydroxyl might also substitute for the sugar hydroxyl on the protein-free end of the normal topoisomerase-DNA covalent intermediate. If the scissile strand –1 T is replaced by a ribouridine in the vaccinia topoisomerase-catalyzed DNA cleavage, the expected covalent intermediate is formed with the active site tyrosyl O4 of the enzyme joined to the 3′-phosphoryl of the uridine. But, the adjacent 2′-hydroxyl group of the uridine can now attack the enzyme-linked phosphorus, freeing the enzyme and forming a cyclic phosphate (Fig. A2-2).[6] Thus vaccinia topoisomerase would act as a nuclease if

5. Wittschieben, J., et al. 1998. *Nucleic Acids Res.* **26:** 490–496.

6. Sekiguchi, J. and Shuman, S. 1997. *Mol. Cell* **1**: 89–97.

there is a –1 U substitution in its normal DNA substrate.[6] Conversion of a nuclease to a topoisomerase-type DNA breakage-rejoining activity by a single amino acid substitution in the enzyme has also been observed. Substitution of Leu43 by a lysine in the sequence-specific endonuclease NaeI, for example, has been shown to convert the enzyme to a DNA breakage-rejoining activity.[7] These examples show that catalysis of DNA breakage and rejoining by a DNA topoisomerase is closely related to the actions of other nucleic acid enzymes.

7. Jo, K. and Topal, M.D. 1995. *Science* **267:** 1817–1820.

Index

Page references followed by f denote figures; page references followed by t denote tables.

T